ENVIRONMENT PAPUA NEW GUINEA

ENVIRONMENT PAPUA NEW GUINEA

Collected Papers Series

Volume No. 1

Edited by
John Rivers
F.L. (Rick) Bein
and
Philip Siaguru

Environmental Research and Management Centre
Papua New Guinea University of Technology
LAE

Originally published in 1999 by:
Environmental Research and Management Centre
Papua New Guinea University of Technology
Lae
Papua New Guinea

ISBN 978-9980-945-74-7

Reprinted with permission in 2012 by:
University of Papua New Guinea Press & Bookshop
P. O. Box 413
UNIVERSITY PO
NCD
Papua New Guinea
Tel: 675 + 326 7281 or 326 7675
Email: upngbooks@gmail.com

Environment Papua New Guinea

Environment Papua New Guinea aims to create a forum for Melanesian research with a focus on the contemporary environmental and developmental problems of the region. The editors welcome contributions which address the environmental and social aspects of regional development and conservation as well as conceptually innovative studies that place regional problems within the wider world perspective. ***Environment Papua New Guinea*** aims to be interdisciplinary in character with an emphasis on exploratory and innovative work which draws on different disciplines and addresses core issues in Melanesian environmental, social and developmental studies.

Editorial and business correspondence should be addressed to:

The Editors, Environment Papua New Guinea, Environmental Research and Management Centre, PO Box 39, Unitech PO, Lae, Papua New Guinea.
Telephone (675) 473 4686/473 4233. Fax: (675) 475 7505.

Notes for Contributors

Environment Papua New Guinea publishes collections of articles on Papua New Guinea and Melanesian environment and development issues. Submissions that fall within this scope are welcomed.

Manuscripts should be sent to The Editors, *Environment Papua New Guinea*, Environmental Research and Management Centre, PO Box 39, Unitech PO, Lae, Papua New Guinea. If a manuscript has been prepared by computer methods then we welcome receiving a diskette copy (*Word* for *Windows NT* format preferred). It is acceptable to submit typewritten or handwritten material but this must be clear and legible. All submissions will be independently refereed. Shorter notes, which may refer to articles appearing in earlier volumes, will also be published, although it will be policy to refer these both to referees and the original author. Letters to the editors are welcomed. Authors should note that articles exceeding 5000 words in length will not normally be published. Articles should be typed on one side of the paper, double spaced and with ample margins. There should be a title page bearing the title of the contribution and name(s) of the author(s) and the full postal address of the author who will check proofs and receive correspondence and offprints. All pages should be numbered.

An abstract of 50-150 words should be included. Footnotes to the text should be avoided.

Tables and captions must be typed out on separate sheets and not included as part of the text. Tables and figures should be numbered by Arabic numerals. The approximate position of tables and figures should be indicated in the manuscript. Captions should include keys to symbols and should be grouped together.

Figures and line drawings should be of a suitable quality for printing and will not normally be redrawn by the publisher.

References should be indicated in the typescript by giving the author's name, with the year of publication in parentheses. If several papers by the same author and from the same year are cited, (a), (b), (c), etc. should be put after the year of publication. Titles of journals should not be abbreviated. The references should be listed in full at the end of the paper in the following standard form:

Books, monographs:
Serpenti, L M (1965) *Cultivators in the Swamps: Social Structure and Horticulture in a New Guinea Society*. Assen: Van Gorcum.

Multi-author volumes:
Pearl, Mary C (1994) "Local Initiatives and the Rewards for Biodiversity Conservation: Crater Mountain Wildlife Management Area, Papua New Guinea." In *Natural Connections: Perspectives in Community-Based Conservation*, edited by David Western, R Michael Wright, and Shirley Strum. Washington DC: Island Press.

Journal articles:
Kirsch, Stuart (1997) "Regional Dynamics and Conservation in Papua New Guinea: the Lakekamu River Basin Project." *The Contemporary Pacific* 9:97-120.

Proofs will be sent to authors. They should be corrected and returned to the publisher within three days. Major alterations or revisions cannot be accepted.

Articles accepted for publication become the copyright of the Environmental Research and Management Centre of the Papua New Guinea University of Technology.

Contributors

D.A. Askin works for the Baptist Church of Aotearoa New Zealand aid programme.

R.A. Curley is a Principal Technical Instructor in the Department of Surveying and Land Studies at the University of Technology, Lae.

F.L. (Rick) Bein is Director of the Environmental Research and Management Centre at the University of Technology, Lae and Professor of Geography at Indiana University Purdue University, Indianapolis.

Peter Clark is Director of the Rain Forest Habitat Centre at the University of Technology, Lae.

I. McL. Grant is veterinary surgeon and works with ACNARS, Lae.

Hartmut Holzknecht is a Visiting Fellow at the Department of Geography, Research School of Asia Pacific Studies, Australian National University, Canberra.

Alphonse Kambu is a student of environmental law at Chiba University, Japan.

Alan R. Quartermain works with ACNARS, Lae, and advises on agricultural research in Papua New Guinea.

John Rivers teaches development studies at the PNG University of Technology, Lae.

Fritz Robinson works with ACNARS, Lae, and advises on agricultural research in Papua New Guinea.

Frieda Siaguru is a Lecturer in the Department of Accountancy and Business Studies at the PNG University of Technology, Lae.

Philip Siaguru is Pro Vice-Chancellor (Academic), Associate Professor and former Head of Forestry at the University of Technology, Lae.

G.K.N.S. Subasinghe is a Senior Lecturer in the Department of Mineral Chemistry at Murdoch University, Perth.

Brian Young is Professor of Civil Engineering at the PNG University of Technology, Lae.

Acknowledgements

The Papua New Guinea Environmental Research and Management Centre was initiated at the PNG University of Technology in 1994 to promote the discussion of the environment among the academic community and in Papua New Guinea generally. As a university Centre it answers to the Vice-Chancellor and works with academic departments that demonstrate environmental interests. Some of the goals of the Environmental Research and Management Centre are to develop the teaching of environmental concepts both at the post-graduate and under-graduate levels, and to expand knowledge of the environment through interdisciplinary research and consulting. A new publication has been approved for the Environmental Research and Management Centre called *Environment Papua New Guinea*. This will provide a forum for environmental research on Papua New Guinea. With respect to community service the Centre is developing outreach programmes to non-government organisations and to schools and is offering short courses and seminars in a variety of settings in Papua New Guinea. The Centre is also coordinating environmental assessment teams to study the effect of development projects on the physical and human environments. The Centre plans to publish further volumes in this series.

We would like to acknowledge the assistance of the Vice Chancellors, Misty Baloiloi and Dr James Kaiulo, for their support for this publication. Also we would like to thank the Centre's Board of Directors for their support.

The Environmental Research and Management Centre can be contacted at Private Mail Bag, Unitech, Lae, Papua New Guinea. Telephone (675) 473 4686 and fax (675) 475 7534.

Contents

1

Gender and Environment Issues in Papua New Guinea

Frieda Siaguru

ABSTRACT

Women have a highly valued role in traditional Papua New Guinea culture. Traditionally, women in PNG are the primary food producers. These necessities are derived from the land and forests, hence fastening women to environment. This close connection between women, domestic subsistence production and the contribution that women can make to the care of the environment is often overlooked. Contemporary Papua New Guinean society has chosen not to notice this contribution. Studies by Assail (1995), United Nations (1994) and the Papua New Guinea Government have highlighted some of the reasons and shown that women continue to experience the negative impacts on the environment of economic development they share almost none of the economic benefits. As governments formulate policies to promote forestry, mining, fisheries and other investments rural women continue to rely on the land to feed their families. Most of these policies do not reflect the values of women and their dependence on the environment. Women's role in agriculture often clashes with the pressures of sustainable subsistence farming (NFCAP 1994) and other investor influences. The focus of this paper is on the problems facing women created by the industrial policies for national development. It emphasises the continuing dependence of rural Papua New Guinean societies on subsistence production and the long-term need for sustainable rural land use.

Introduction

The traditional women's role in Papua New Guinea is tied steadfastly to subsistence food gardening and traditional domestic practices, which over many generations have become set within the Papua New Guinean cultural framework. All Papua New Guineans have strong spiritual affections for, and ties to, their local environment. But women, with their primary role as household food producers, are closer in many respects to their environment and more directly reliant on the benefits of a well-managed local environment than men. The traditionally important roles of women in Papua New Guinea are connected through customs, folklore and tradition. Women in Papua New Guinea were the primary food producers. Domestic household production was derived from the land and forests, fastening the women more to the environment. The contribution of women is important, but often not mentioned. In spite of government policies calling for gender equity, women continue to be visibly absent in the decision-making process, both within government committees and in traditional village government.

Recent surveys, as discussed in the Annual General Meeting reports of the Papua New Guinea National Council of Women, have pointed out that women's workload, access to land, health and social status is on the decline. The worsening of the situation has come about as a result of the government's policies for increasing forestry investment and the conversion of land for large commercial scale agricultural use as oil palm plantations, beef cattle ranching and peanut farming. Other land use such as mining is also being established at the expense of the household subsistence sector. The change in the pattern of land use favours men and disadvantages women in the pursuit of their traditional occupation as household food providers. Associated with these land use changes has been a rapid rate of urbanisation which has also caused displacement and a deterioration of the cultural practices of family sustenance. Women who were comfortable and self-reliant by tilling the soil for the upkeep of their family now find themselves thrust into a situation where they lack education, knowledge, skills and the confidence that are needed to make the transition from a traditional style of life to an urban style of living.

These cultural changes are traumatic and often cause social disorders on a large scale. Development and sustainable projects at village level must, therefore, in my opinion, encompass all of the following factors.

Development of the self

Just as a baby develops in the womb of the mother, so must any project. It must seek to develop the mental, physical and spiritual dimensions of its participants, and keep the project in balance with the cultural, political, social and economic relationship factors that are all involved. Failure in any one of these aspects will almost certainly mean the demise of others.

Vision

Our tribal ancestors governed with vision. They knew and respected nature and gave it values equivalent to their own existence. These visions built traditional societies that existed harmoniously within themselves and their environment until 'Papua New Guinea discovered the first explorer.'

Sustainable Development

To be sustainable over the long run a concept or a project must be kept going. It must have the necessary environmental checks and balances built in from its inception.

Knowledge

This comes with education and free and unconditional sharing of information. Knowledge should be free of gender discrimination and should draw on the best available human resources.

This chapter advocates the use of the four principles mentioned above. These principles, from my experience and observation, are factors that have been over-looked and often misinterpreted. The paper will describe gender equity as it was and is and will endeavour to visualise the concept of gender participation as Papua New Guinea moves into the 21st Century. It is, in a nutshell, arguing for the need to listen to the silent voice of a gender that values and ranks care for the environment as a priority. Our reason for this care is simple: we want a future that is progressive for the children.

Rationale

Women generally are discriminated against in Papua New Guinea. Our views are usually considered to be of secondary and lesser importance to those of men. While the Constitution supports equality of rights and government policies encourage equal participation in nation building, real life situations are contrary to these policies. This is especially so in the urban villages and settlements as well as in rural communities. Women's value, however, remains as the most important asset in these underprivileged communities. But the significance of women's contribution for better communities continues to be ignored by the government and its policy-makers. Reports compiled by UNDP (1975) show that, while the Government of Papua New Guinea recognises the role of women in society and assisted in the establishment of the National Council of Women in 1975, no real efforts were made then and since to translate women's value and input into national development policy.

Let us now look more closely at the rural environment and the connection of village women to the environment. A typical village mother wakes up at about 6 a.m. Her first concern is breakfast. After the household has had breakfast and the children go off to school with roasted root crop or banana for lunch and the husband sets off for 'meetings', the mother packs her tools and goes off to the food garden. There she spends all her day, tilling the soil and planting food crops. At the end of the day, she hurries back with some harvested food for the family dinner. She will have in her *bilum* bag vegetables, some root crops and almost certainly a bundle of firewood on her head for cooking the evening meal. After the evening meal the children will go off to sleep and the men go to the men's house. The woman then washes her dishes or at least puts away the left over food and prepares the kitchen for the next day's breakfast. Her day ends at about 9 pm. She takes an hour's break for a shower perhaps, and then takes some time to attend to other household chores. Then she goes to bed.

Children often fall ill. When they do so the mother will immediately suspend her daily schedule and attend to the health of her children. If a modern clinic or medicine is not available, the mother has to rely on traditional herbal medicine. Knowledge of traditional medicine is passed down from mother to daughter. Often, there will be a specialist (in most cases a woman) in the village who knows the traditional medicinal plants. These knowledgable people are consulted and the correct traditional medicine is prescribed and applied. Papua New Guinea village communities have lived off and with the rainforest for generations and form an integral part of the complex forest ecosystem. Often their interaction with the forest is mutual rather than antagonistic. Village people have learned to live in harmony with their environment and with a lifestyle that uses the forest while respecting its limited fragile carrying capacity. Hence, any damage to the environment system by clear felling or making a commercial plantation destroys forest diversity and so damages forest people.

I remember overhearing a conversation between a international marine biologist and a waitress at a coastal resort in Papua New Guinea. The marine biologist, after painstakingly describing his observations of the local marine life and commenting seriously about the fact that the sea water level was rising, turned to the local waitress and asked, "What will you do, if the sea water level rises and submerges your home". The waitress replied instantly, "I will pack my things and my family and move up to the mountains and build a new home". The scientist was dumbfounded, since he was expecting a response that would be more in his line of thinking, one of taming, exploiting and controlling nature. This anecdote illustrates the simplicity and the respect we have for mother nature. We do not seek to dominate nature, but instead desire to live harmoniously with it.

School fees are often the largest single expense faced by a Papua New Guinean family. Village mothers make an important contribution to raising and paying school fees. Extra crops are planted and surplus food from their food gardens is sold in the town markets to raise money for school fees. This is often a long and laborious process that goes on continuously throughout the year. Often some 50 per cent of the cash raised must be used to pay the cost of transporting the food to the urban market.

Women's knowledge of plants and their uses is often encyclopaedic. It is part of every village woman's education to know which plants are medicinal, which plants are for decoration for special village occasions and which plants are out of bounds and sacred. The mother knows which special plants are used to cover babies' with when they are born, to clothe them when they are initiated and to dress and decorate them when they are to be presented to their future spouses. The mother's connection to the environment is at the basis of the families livelihood. This connection makes women the first to feel the loss when the environment is disturbed or destroyed. Women feel the loss more than men. Not only must she feed the children but the husband as well. A woman becomes insecure and fragile if her local dependence on the environment is removed or broken. She needs her traditional lifestyle as a companion. Bosse (1994) argued that, due to such expectations, the new generation of girls were forced to develop self-reliance very early. They begin experimenting with femininity projects also. The maturity process concerns the search for guidance on who one is and what it means to be a female. Bosse (1994) further shows that these girls become aware at a very early age that there is no ready-made pattern for female identity that can be followed. The mothers and females in the village can provide guidance, but the sole responsibility of survival and charting a path in the future lies with her and, eventually, her influence in her family.

Development and Gender

Amidst all the efforts put in by women to sustain their families and themselves, men in Papúa New Guinea make all the decisions for development in the villages. Men decide on forestry projects, agricultural projects, mining projects and other major changes to land use. Men decide on royalties and spend all their time talking about how much money different development projects will bring to rural areas and to their communities.

Studies undertaken by Australian International Development Assistance Bureau (1993b) emphasised that traditional farming systems operated with a clear distinction between men's and women's labour. Typically, men undertook such heavy work as clearing the forest, building fences and planting crops that have cultural significance for them. The remainder of the work is done by women, including planting, weeding and cultivation, and harvesting and cooking. The report describes the pattern of men's and women's activities in the following terms: "Men had burst of activity followed by a lengthy leisure period, whilst the women had a life of constant work". Another study by Overfield (1993) of coffee cash cropping shows that women are doing half of the work but their men appropriate most of the income. Women are disadvantaged in most aspects of rural life. They receive lower wages for agricultural labour, earn less than men from the sale of cash crops, rarely become owners of settlement blocks or hold customary land in their name. Allen (1993) pointed out that in the highlands, women produce and market most of the food crops and raise most of the pigs that give status to men. Increased trading in traditional food items, although creating economic competition among women, is also increasing the possibilities for participation in income generating activities. But these increases in trade link women even more to cultivation and agriculture. Small holder farmers collectively can produce sufficient food to support the growing Papua New Guinea population, even though some land has been allocated permanently to cash crops (AIDAB, 1993b).

AIDAB (1993b) in a survey conducted on 574,000 householders, reported that about 75 per cent of the people of Papua New Guinea depend on a mixture of gardening for food and production of tree crops, root crops or vegetables for local domestic sale. Another 8.5 per cent live in remote areas and garden exclusively for subsistence while 4.5 per cent work exclusively in modern cash based farming. The formal cash agriculture sector is dominated by the tree crops, coffee, cocoa and oil palm. Historically, copra dominated cash cropping in the coastal areas but has now declined and cocoa has increased in importance. Cash cropping of vegetables is done by about 270,000 households and coffee and banana by 246,000 and 234,000 households, respectively. The full results of this AIDAB survey are set out in the following table. (Table 1.1)

Table 1.1: Number of households growing cash crops, 1990 ('000)

Vegetables	270
Coffee	246
Banana	234
Betel nut	134
Coconut	111
Cocoa	93
Spices	52
Other crops	31
Rubber	8
Oil Palm	8

Source : Papua New Guinea, Department of Agriculture and Livestock, *Rural Statistics Bulletin 1993/2, Port Moresby.*

Table 1.1 confirms two factors. The first is the fact that Papua New Guinea households depend on a system of domestic agricultural production. As long as this situation continues, women will stay at the forefront and deepen their connection to the environment. Papua New Guinea is a country where people do not have access to a government funded social security system. This is a country where, for a long time yet, agriculture will remain the focus and thrust of government policies for integrated rural sustainable development. This is a country where survival for many people requires the skills to cultivate the land. Hence it enhances the women's connection. Women have been cultivating the land and building a cultural status for their sons and husbands since time immemorial. The second fact establishes the role of women in the over-arching national objectives of Papua New Guinea. We have silently over the years proven our worth and we must now, as equals, share collectively the task of nation-building.

Resource exploitation and land use projects do impact on women more than men. A study conducted by Vatasan, Siaguru and Simaga (1995) in West New Britain Province (where about 60 per cent of all logging in Papua New Guinea is conducted), further highlights this female connection to their environment. In the study 100 per cent of the women interviewed expressed that any community land use based project should aim at making the people, especially mothers, self-reliant. They wanted coffee peeling machines, sawing machines, basic education in improving their garden produce and, the strongest request, to reduce destruction to the rivers, water ways and hill slopes. This request ties into community projects that can turn over cash over the long period of time and projects that women can manage and direct at the village level. The study further showed that women were more interested in projects that will help them earn steady income for their household while their menfolk were more interested in the monetary returns and royalty moneys that can be raised from logging projects. Women also wanted the investors to be required to spend money on environment rehabilitation, while men demanded compensation for environmental damages to be paid in cash. In Amio village, located south of Kimbe, women wept openly when a ship that had called in to pick up logs discharged oil in the bay and polluted the coral reefs and local fishing grounds for a radius of about 5 kilometres. While the women wanted either the state or the operator to clear up the mess, the men were busy demanding cash compensation.

Women and the Environment Today

How have women in Papua New Guinea fared so far? The trend is changing. We do not condemn and blame our menfolk for their lack of concern. It is simple their priorities are different to ours. Since they cannot speak up for our cause, we must take that lead ourselves. We must stand up and tell it ourselves.

At the 1994 and 1995 annual general meetings of the Papua New Guinea National Council of Women, there were major concerns raised about the destruction and degradation of the environment. Women from the Papuan Region highlighted pollution from the OK Tedi Mine. Since no proper tailings dam for the containment of mine waste has been built all of the waste is discharged directly into the Ok Tedi and Fly River systems which provide breeding grounds for barramundi, prawns, crabs, mud fish and other forms of fresh water life which are a major food resource in the region. The mine discharge has also affected and destabilised the quality and quantity of sago, which is a staple food item for people living along the Fly River. Logging projects in Oro, Milne Bay and Central provinces are also creating pollution and social problems. Also from Oro it was reported that careless discharge of oil associated with oil palm production has some toxic effects and leads to a declining quality of food garden production in the affected area. The reports also highlighted major ecological destruction to the environment with the construction and laying of the pipelines for transporting oil and gas from Lake Kutubu region to the Gulf of Papua.

From 1991 to 1993 intensive logging was conducted on Siassi Island in Morobe Province. Through an intensive study mentioned in EMW (1995), it was concluded:

- The majority of the population was excluded from the initial discussions and contract agreements.
- Supervision of log exports and timber felling is inadequate.
- Deforestation causes erosion which leads to nutrient loss and hence agricultural potential declines.
- Acquisition of timber is conducted without concern and respect for nature.
- Within the vicinity of loggers' camps, it was noticed that the beaches were strewn with batteries, tins, broken bottles and other human discharges.
- Reefs were destroyed, hence fishing grounds lost when coral was quarried to surface logging roads.
- Drinking water was destroyed when crude oil and fuel were discharged into rivers and streams. Hence, women have to walk long distances to fetch water for drinking and cooking.
- Anti-social activities increased.
- Loss of cultural sites and taboo plants.
- Destruction of the environment has resulted in an increase in tropical diseases.

EMW (1995), reported Sihaila Talau, Chairperson of the Women's Committee on Siassi, as saying, "When the company came, women were not allowed to participate in the discussion as to whether or not we wanted deforestation. If we had been present the whole affair would have taken a different turn and the timber companies would not have got to Siassi. Because we use many rainforest plants for food, medicine and the manufacture of other products, we know how important an intact forest is to us. A lot of men have forgotten how essential the rainforest is for Siassi because they only have one thing on their minds: money! And as soon as it's there they spend it carelessly or drink their way through it."

Almost all indigenous landowners care deeply about the state of their land. The following are some of the major environmental concerns raised on a large scale in Papua New Guinea. The Japa River pollution on Bougainville (a matriarchal society) raised considerable local feelings on Bougainville before the rebellion began in 1988. The other controversy has been over the OK Tedi copper mine dumping its waste down the Ok Tedi and Fly rivers, which ultimately ends up in the Torres Strait. East of the Fly delta is the terminal where crude oil piped from the Kutubu field is loaded into tankers. This takes place where there are large barramundi and prawn fisheries and not so far the beginning of the Great Barrier Reef. The Misima mine is close to a deep ocean trench, EMW (1995) reports that mine wastes are dumped off the island into the sea. The Lihir mine sits on a projecting shelf from the island, hence wastes will have to be piped out to a deep trench or shipped out.

The above projects have a direct affect on women since women continue to be the domestic producers and providers. I recently conducted a simple survey on women's input into family management and dependence on the environment. The survey was done to gauge a national perspective. Fifty women currently living in the village were interviewed. The population surveyed was small, but it did, however, gave a good spread of perspective. The survey was constrained by the fact that I did not travel to the different provinces, but I located mothers who are from different provinces currently on holiday and living with relatives at the University of Technology. Part one of the survey concentrated on family sustenance. This was to establish whether rural communities are supplementing their sustenance from cash earned from a salary or continue to live on subsistence gardening. Part two of the survey investigated family maintenance and management. The survey wanted to establish if the traditional roles of women as house-keeper and provider differed between the urban and the rural context.

The results suggest that rural communities still rely on their sustenance from subsistence food gardening. Families where only one family member is earning an income or salary continue to supplement their sustenance through food gardens. Family upkeep and maintenance showed that 100 per cent of the women indicated that they continue to provide the stability in the homes. However, 90 per cent of the women interviewed indicated that overall decisions in family matters and initiation of discussions on matters regarding family upkeep are made by males. The traditional roles of gardening, laundry, child care and food provider remain steadfastly with women. The survey also strongly shows that the above situation is also true for families living in urban settlements, urban villages and rural communities. This suggests that a salaried working woman continues to provide the same attention and input to family maintenance and upkeep as does a woman working in the village. However, the rural and urban village women depend more on subsistence gardening for their family upkeep since they have no supplementary cash income. Urban working women supplement their families' upkeep with their income. Therefore, all the three categories of women surveyed become very vulnerable when the environment is disturbed or mismanaged. Hence, one can see the deep connection women have with the environment. It is traditional, cultural, and it forms the very basic fabric by which the woman maintains the family.

Women and the Environment Tomorrow

Women have to continue to speak out. We must gain equal attention in our efforts to raise the hitherto silent, but persistent, voices of concern and conscience within us. The annual general meeting of the Papua New Guinea National Council of Women in 1994 passed summary recommendations that called for the economic empowerment of women and the monitoring of the Global Platform for Action. It was clearly expressed that women cannot be truly free unless they are financially independent or have economic power. This view was also expressed by Ronnie Alexander, Professor of International Relations at Kobe University (Japan). Professor Alexander was a speaker at the conference "No Quiet Ocean — No Peaceful Pacific" held in Germany in 1992 and organised by the Evangelical-Lutheran and Catholic Churches of Bavaria.[1] Professor Alexander pointed out that since the end of the cold war endogenous issues in the Pacific have increased in importance and mentioned three as most obvious, (i) political independence, (ii) economic development and environmental sustainability, and (iii) resource conservation and environmental sustainability. Mr. Charles Lepani (then Papua New Guinea Ambassador in Brussels) speaking at the same conference pointed out that the majority of Pacific Islanders are dependent on rural subsistence production. Alex Maun, the spokesman for the affected OK Tedi/Fly River Landowners' Association, remarked that "the Government of Papua New Guinea and OK Tedi Mining Ltd. have violated the law of the Constitution, of human rights of indigenous people and customary rights."

I have highlighted the issues affecting women living in mining affected areas. But the same situation exists in forestry, fisheries, agriculture and other general land-use projects. Women continue to have a silent voice. But it is my strong belief that this is changing. Women are now becoming more alert and determined to conserve and sustain the natural environment that is the source of our livelihood. Indeed, we should be strong and say how women fare in all the investment development programmes that are encouraged by the Papua New Guinea government. I believe that women are becoming radical modernisers. They have the ability to rework old traditions in Papua New Guinea and integrate them into the modern economic, political, and cultural systems (Bosse, 1994). The study by Bosse (1994) shows that women are taking up dialogue on what Papua New Guinea is today and what it will one day become in the future. And more importantly, what Papua New Guinea should not be allowed to become.

In my view, the character change in modern women occurred due to environmental circumstances created by investors operating in PNG and the policies of the Papua New Guinea government. Dialogue of this new kind was not encouraged in traditional village society, perhaps due to customary obligations. However, both sexes knew the hidden strengths of the culture of the opposite sex. The sexes complemented one another and together they regulated the continuance of the group (the only difference was that the women had a silent voice). That, to me, is the foundation women should stand upon with our menfolk and mutually take Papua New Guinea into the next decade, but this time, our voices must be heard, particularly on issues that will affect our environment and our livelihood.

Conclusion

Although environmental impact studies are a compulsory requirement by the state, few investors bother furnishing such studies. Since baseline environmental information is usually not available the end result is that many people continue to suffer from the bad effects of investment projects. Sustainability in the traditional context depended largely, perhaps, on situations available then. We have to be truthful, almost all models of sustainability are historical, since the style of living contained in it ideas and practices that could allow the environment to adjust over time. For example, in agricultural practice the fallow period was sufficient to allow abundant produce in food gardens on successive rotational cultivation. This was the concept of the *gut taim*,[2] as seen through the eyes of our ancestors. But this was also a time when population was low, natural mortality was high, village communities were small, tribal warfare was endemic and traditional methods of hunting and cultivating were primitive and inefficient (this is not to say that our style of living and our ancestors were primitive in a derogatory manner, but to highlight the true perspective of sustainability). Within that context, our lifestyle conformed to the rule of mother nature and, therefore, was sustainable. I wish to reinforce the fact that people and earth are interdependent. Through this people were able to form the factor of sustainability.

Hence, it is my strongest view that current practices of sustainable development must take the same line. All of the prevailing pressures influencing the sustainability of the human/ environment relation must be taken into account. In my introductory remarks, I made reference to four major points, development of the self, vision, sustainable development, and knowledge. Similar views have been expressed more specifically in relation to logging issues and land use. Siaguru (1994) highlighted sustainable management involving a whole eco-unit. Such eco-units must consider: (i) displacement of villagers, (ii) flora and fauna extinction, (iii) imbalance of village existence, (iv) reduction of economic potential, and

(v) environmental degradation. Omission of any one part of the above will almost certainly mean a failure in any sustainable management model.

Much published literature has highlighted time and again the need to identify people with the environment more than anything else. Development investment that leads either directly or indirectly to deforestation is having a serious impact on plant and animal species. By far the most direct and immediate and long-lasting impacts are on the native peoples. Within these native peoples, the group that feels the pain more are the women. Women living in rural and urban village as well as urban settlements in Papua New Guinea bear the real force of the social and ecological costs of investment since the first and greatest loss to them is the use of the environment. Our men folk may be lucky to get some rewards while the full returns on these investment go to a relatively small minority in Papua New Guinea and the world. These loses are significant since there is also a major cultural loss because the indigenous villagers will also lose with the disappearance of the environment the traditional knowledge of the forest and its sustainable techniques. Hence, we must all work collectively to make PNG become responsible and manage the environment in a way that will sustain itself and its people, which will inevitably make the world a better place to live in and grow our families collectively. Indeed, it often makes me ponder the words of Chief Seàttle, "We don't own the land, the land owns us" which is the tail of what we say here in PNG, "we own the land."

NOTES

1. This conference eventuated as a result of the Starnberg Report and its critique of environmental degradation in Papua New Guinea. In October 1991 the Melanesian Institute published a report on "Development and Environment". This report was the result of preliminary research done by Dr. Kreye and Professor Castell of the Starnberg Institute. The study looked at the environmental impact of the Ok Tedi mine and the sustainability of social options along the Fly river following the commencement of mining operations by Ok Tedi Mining Ltd.

2. The concept of *gut taim* (good time) is often mentioned by the old people in the village. It is told in legends and traditional stories around the fireplace in the village and in the *haus man* (men's house). It tells of the time when meat was plentiful, rivers were flowing with clean, clear water and generally when life was peaceful.

REFERENCES

AIDAB (1993 a). Papua New Guinea - Australia Development Cooperation Program: Social Impact and Gender Analysis, Renewable Resources Sector Study. Waigani.

AIDAB (1993 b). "The Papua New Guinea Economy." *Prospects for Sectoral Development & Broad Based Growth.* International Issues No. 30.

Allen, B. (1993). "Issues for the Sustainability of Agriculture Systems in Papua New Guinea, 1990 - 2015." Canberra (unpublished paper).

AUSAID (1995). "Women in Development." A USAID project document on gender participation in Papua New Guinea. Australian High Commission. Waigani.

Bosse, H. (1994). "Becoming a Papua New Guinean." *A report of a Sociologist's and Group Analyst's Research with Students at Passam National High School.* National Research Institute Discussion Paper No. 78. Port Moresby: National research Institute.

Barber, K. (1993).: "The Informal Sector & Household Reproduction in Papua New Guinea." National Research Institute Discussion Paper No. 71. Port Moresby. National research Institute.

Association of Protestant Churches and Missions in Germany, EMW (1995). "Our trees and all the wildlife have gone." *The destruction of the tropical rain forests, media conflicts, and the struggles for a holistic human development in the Pacific region.* EMW, Hamburg, Germany.

National Forestry and Conservation Action Programme (1994). Review of the NFCAP programme. A review programme funded by UNDP. Port Moresby.

United Nations (1994). *Yumi Wankain.* Report of the United Nations Joint Inter-Agency Mission to Papua New Guinea on Sustainable Development. Port Moresby: UNDP.

Overfield, D. (1993). *The Economics of Smallholder Coffee Production, Analysis, Policy and Future Research directions.* Goroka: Papua New Guinea Coffee Industry Cooperative.

Papua New Guinea National Council of Women (1994). Annual General Meeting Report. Edited by U. Wainetti.

Siaguru, P. (1994). "Utilisation of PNG Forests in Harmony with their Conservation," in *Proceedings of the 1994 Obihiro Asia and the Pacific Seminar on Education for Rural Development (OASERD).* Obihiro University of Agriculture and Veterinary Medicine. Hokkaido. Japan.

Vatasan, G., Siaguru, P. and Simaga, J. (1995). "Environmental Audit for Stettin Bay Lumber Company." (unpublished manuscript).

2

Pollution Aspects of Mercury Used in Alluvial Gold Processing

G.K.N.S. Subasinghe

ABSTRACT

Mercury is known to recover fine gold particles very effectively through amalgamation. It has also brought about considerable environmental pollution and health hazards to small-scale mining communities. This chapter highlights the malpractices in the use of mercury by small-scale miners and the harmful effects resulting from such practices. It also discusses possible ways of reducing mercury pollution in the Papua New Guinea context.

Introduction

Mercury pollution has been recognised as the most disastrous aspect of the small-scale gold mining industry around the world. In Latin America, over one million people are estimated to be involved in such operations (Hentschel and Priester 1991), and invariably use mercury for "amalgamation" in order to recover the fine gold particles. The problem of mercury pollution arises from the primitive mining techniques and the ignorance of the health hazard associated with such pollution. Particularly in Brazil, mercury pollution has become a large social problem. It is estimated that up to 200 tonnes of mercury may be used per annum by small-scale miners and even in Papua New Guinea where over 4,000 miners are involved in such operations (Blowers 1988), the extent of mercury use can be over 10-15 tonnes a year. Remedial measures taken to eliminate the mercury pollution such as the legislation of mercury bans in other countries have been largely ineffective due to ineffective monitoring and illegal mining operations. The aim of this paper is to describe the type of mercury pollution that can result from such operations and to discuss the strategies that can reduce the release of mercury into the environment and alleviate the risk to the people involved.

The Use of Mercury in Small-scale Gold Mining Operations

The use of mercury has become popular among small-scale miners because amalgamation is known to efficiently extract fine gold particles from concentrates obtained by panning or sluicing operations. Gold alloys with mercury to form an amalgam from which the gold can subsequently be separated by evaporating the mercury. The simplicity of the technique, low investment cost and its comparatively high gold recovery rate have made the mercury amalgam method an integral part of the small-scale gold mining operations.

In Papua New Guinea most of the gold deposits worked by small-scale miners are alluvial deposits in which the gold particles are liberated from the gangue particles. It is customary to use riffled sluice-boxes to recover the liberated gold particles. However, some of the gold particles, particularly the fine gold, does not settle out in riffle compartments and flows over to the discarded tailings (Subasinghe 1993, Kelly, Subasinghe and Walsh 1995). In the hope of trapping some of these fine gold particles alluvial miners frequently place some mercury in between the riffle compartments. Due to agitation and also subsequent mishandling during the withdrawal of mercury from the sluice-boxes, approximately 5 - 30 per cent of the mercury may be spilt or discharged into the river system.

The most dangerous practice adopted by the miners is the gold recovery process from the amalgams. The gold is recovered by evaporating the mercury from amalgam over an open fire. This process is commonly known as "cooking". The mercury vapour, which includes fine globules, is partly inhaled while the remainder is released to the atmosphere, which returns as part of the 'mercury-cycle' described below. Methods introduced to avoid the practice of releasing mercury into the atmosphere and which can reduce the mercury loss to less than 0.1 per cent, are available but have not been popular among miners, due to the discolouring effect on the gold after distillation in a retort. The discolouration is caused by Fe and As compounds and results in a lower price being offered by gold buyers for the product. One such device is the "Retort" which evaporates the mercury in a closed chamber and recovers it by condensing the vapour with cooling water.

The Toxicology of Mercury

Mercury poisoning drew world attention in 1953 after it was reported that a large number of people living in the Minamata Bay area in Japan developed symptoms of a disease which affected their central nervous system after consuming fish. The fish in the bay were contaminated with methylmercury as a result of mercury released into the bay by the Chisso Corporation, a chemical company operating on the shores of the bay. The mercury poisoning was responsible for a variety of clinical symptoms which included speech impediments, failure of muscular co-ordination, contraction of visual fields in the eye, disturbance in smooth eyeball movements, central hearing loss and unbalance of body. This disease is now commonly known as the *Minamata Disease*.

The toxicity of mercury depends on the oxidation state of the mercury and the type of compound it is present in. Metallic mercury (liquid and vapour) exists as the zero oxidation state and on inhalation is oxidised in the body. In inorganic compounds, mercury exists as Hg^{+} or Hg^{++} ions. The mercuric ion, Hg^{++} can form a number of organomercurial compounds. Any form of mercury has been proved to be toxic to humans.

Metallic mercury is not easily absorbed by the digestive organs but may be absorbed through the skin to some extent. However, it has a rather higher vapour pressure and a high evaporation rate. Even at room temperature, its vapour pressure is 1.3 x 10^{-3} mm Hg which is equivalent to 15 mg/m^3. This concentration exceeds by 150 times the threshold limit stipulated by the World Health Organisation (WHO) of 0.1 mg/m^3. At higher temperatures such as those encountered in "cooking," the vapour pressure is enormously high and may reach 1 atmosphere at 357°C.

Of the mercury vapour inhaled through the lungs, it has been reported that more than 70 per cent enters the body and gets into the kidneys and the brain via the blood stream. It is then, with a half life of approximately two months, excreted from the body. The mercury vapour absorbed in the blood exists as monatomic gas and is eventually oxidised to Hg^{++} in the red cells, but some part of the absorbed mercury can cross the blood-brain and placental barriers before oxidation. Consequently, the inhaled mercury can also affect the brain as well. The toxic effect of mercury can be attributed to the Hg^{++} ions. The inhalation of mercury vapour results in vomiting, gastroenteritis, complaints of the kidney and urinary tract, ulcerations in the gums, and extreme light sensitivity or 'photophobia'. If inhaled over a long period, mercury can cause chronic mercurial poisoning resulting in kidney ulcerations, HgS deposition in the body, speech disturbances and loss of concentration. Insoluble inorganic compounds are hardly absorbed by the human body through the digestive organs and are mostly excreted. But it has been suggested that these mercury compounds can also be absorbed through the skin to some extent.

Organomercuric compounds are highly volatile and toxic. When taken in with food they are easily absorbed by the blood. Particularly, Methyl mercury, CH_3Hg^{+} does not decompose into inorganic mercury in the body, but combines with organic ligands into stable states, which enables mercury to cross the blood-brain and placental barriers. It is then accumulated in the body and causes damage to the brain, especially the central nervous system. Furthermore, in such situations, a baby in the womb is more prone to be affected adversely than the mother herself, and in many cases the baby may end up severely handicapped (Nordberg 1976, Kitamura 1976, Japanese Chemical Society 1977, Suzuki, Imura and Clarkson, 1991, Nakanishi 1992, Nakanishi, Konuma and Gamo 1994).

The mercury released to the environment undergoes microbial anaerobic conversions in the rivers and in soil which leads to methylation of the mercury. The mercury vapour released into the atmosphere is also oxidised into Hg^{++}, due to the influence of ozone, air humidity and UV radiation and eventually comes down by precipitation. The transformation of mercury during the global cycle is shown in Figure 2.1. There is a consensus that methylation of inorganic mercury does not take place in fish or the human body. Although there has been no clear verification of a noticeable methylation of inorganic mercury even in the environment, there have been remarkable increases in the methylmercury content in fish in areas where there is no possible release of artificially produced methylmercury. It has also been suggested that the tropical climate may help the methylation reaction of inorganic mercury in the environment (Hentschel and Priester 1991).

The Mercury Cycle

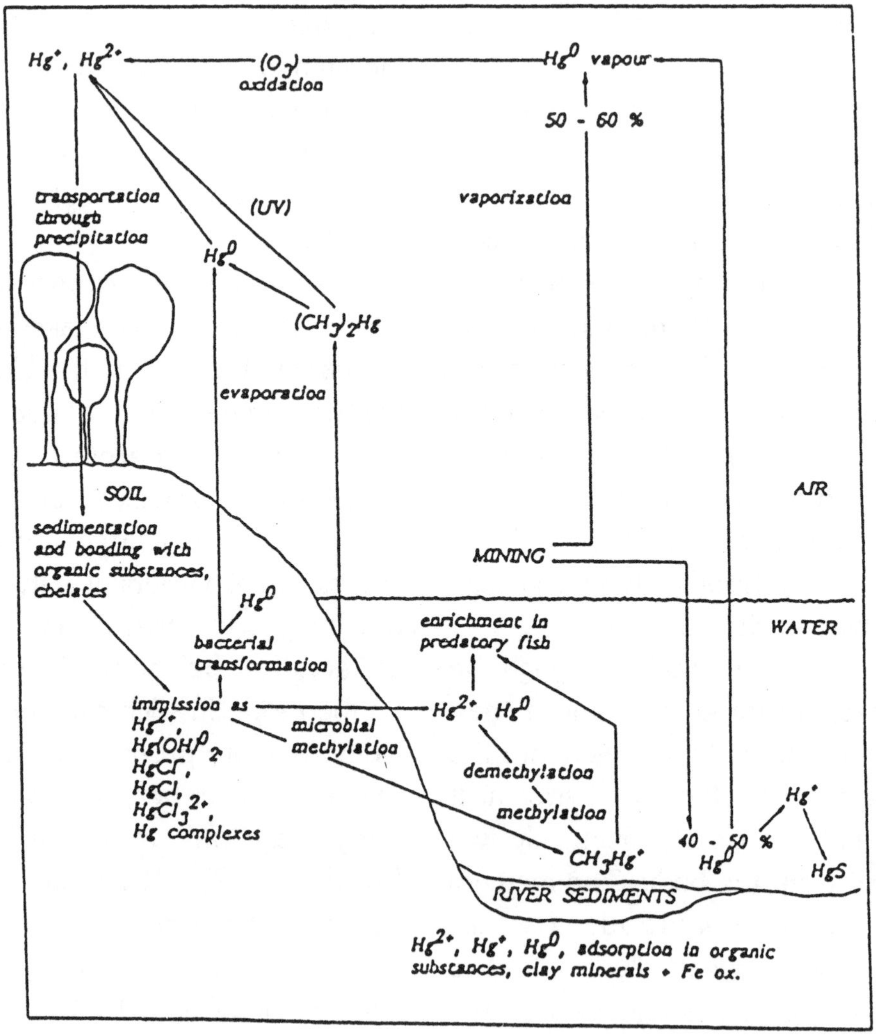

Figure 2.1 Mercury cycle and emission by gold mining activities (after Hentschel and Priester 1991)

The organic and inorganic compounds in different forms enter the human body either by inhalation, drinking water or through food. Of these, the intake of methylmercury from fish is the most important because it is well known that fish can accumulate mercury many times more than the environmental concentration. It is also reported that most of the mercury in fish is in the form of methylmercury (Japanese Chemical Society 1977).

Mercury Pollution by Small-scale Gold Mining in Papua New Guinea

In Papua New Guinea, even though it is known that mercury has been in use in gold mining operation since the turn of the century, the extent of mercury pollution, its distribution and effects on the surrounding eco-systems and humans has not been studied adequately. However, Mallard and Hugman (1996) have reported some results on the impact of mercury on the Bulolo River Valley which has been the centre of gold dredging operations in the past. In Table 2.1, their results are compared with those of some investigations of the Brazilian ecosystems and WHO limits and industrial threshold limits (ITL) for mercury compounds (Hentschel and Priester 1991). In Brazil, it has been found that mercury contamination is not confined to the local areas where mining takes place and there can be a significant increase in mercury levels within a radius of about 200 km around the mining areas. Table 2.2 shows the levels of mercury measured in humans in such regions and those in Wau/Bulolo area of Papua New Guinea.

The data shown in Table 2.1 does not show alarming levels of mercury pollution in the environment of Bulolo River Valley area. However, Table 2.2 shows that the mercury levels in humans engaged in small-scale mining activities can be at alarmingly high levels due to careless handling of mercury. The author has witnessed very dangerous practices of handling mercury by small-scale miners, including direct inhalation of mercury vapour during the "cooking" process and careless and accidental spillages of mercury into the river during the panning and sluicing operations. Considering that "cooking" is usually carried out in close proximity to the dwellings, and in the presence of pregnant women and children, the potential health hazard to these communities cannot be ignored.

Considering that, from Minamata bay methylmercury dispersed to the Shiramui sea until 1968, over 13 years after the dumping of mercury into the bay was stopped, and that the mercury concentration in hair of residents around this area was 10 to 20 times higher than that of people in non-polluted areas, there is some reasonable cause for alarm for the people who live in downstream areas close to the Huon Gulf area and in the city of Lae.

Strategies to Reduce Mercury Pollution

The following strategies can be adopted for the reduction of mercury pollution caused by small-scale mining operations.

(1) Educate people who handle mercury with regard to the health hazards and the safety measures that can be taken in handling mercury by organising workshops.

(2) Establish a national survey of the mercury use in small-scale gold mining and conduct regular health checks on people in the surrounding areas.

(3) The promotion of the use of retorts would be a good solution in the short-term, as they are capable of recovering most of the mercury which would otherwise escape to the environment. The chances of miners being poisoned by mercury could also be minimised through the establishment of some central facilities in Wau area to alleviate the more dangerous practices such as cooking of amalgams.

(4) Look for other alternatives to amalgamation. In this regard, fundamental studies in the development of a flotation reagent that is capable of selectively extracting gold from the concentrates, has been started at the Department of Mining Engineering at the Papua New Guinea University of Technology.

Table 2.1 Allowable limits and measured values of mercury concentrations in Brazil (Hentschel and Priester 1991) and Wau/Bulolo area in Papua New Guinea (Mallard and Hugman 1996).

	WHO/ITL LIMITS	Brazilian Ecosystem	Wau/Bulolo Area
Drinking water	max 4 μg/l	measured upto 11 μg/	Upto 0.47 μg/g in Bulolo area.
Fish	max 0.5 μg/g fresh weight or max 0.2 mg methyl Hg per person per week or max 0.3 mg inorg. Hg per person per week.	up to 2.7 μg/g fresh weight. upto 3.8 μg/g in fish spawn.	upto 0.92 μg/g in Lae area.
Air	For Hg max 0.1 mg/m³ (ITL), For org. Hg compounds max 0.1 mg/m³ (ITL)	upto 0.3 mg/m³ (in the vicinity of amalgamating plants)	

Table 2.2 Measured and normal mercury level in humans

	Measured in contaminated areas (Hentschel and Priester 1991)	Normal (Hentschel and Priester 1991)	In PNG (Mallard and Hugman 1996)
In blood	upto 175 ppb	10-13 ppb	—
In urine	upto 225 ppb	0.5-8 ppb	upto
In hair	upto 40 μg/g dry weight	0.7-3 ppm	upto 5 ppm

REFERENCES

Blowers M J, (1988). *Handbook of small scale gold mining for Papua New Guinea*. Christchurch: Pacific Resource Publications.

Hentschel T, and Priester M, (1991) "Mercury contamination in developing countries through gold amalgamation in small-scale mining: some processing alternatives", Project Consult GmbH, Koenigstein, Germany.

The Japanese Chemical Society (1977) *Suigin(Mercury)*, Tokyo: Maruzen Pub. Co.

Kelly E.G., Subasinghe GKNS, and Walsh D, (1995) "The sluice box - An Evaluation," *Minerals Engineering*, 8(1): 1397-1425, Pergamon Press.

Kitamura S et al. (1976), *Suigin (Mercury)*, Tokyo: Kodansha Pub., Tokyo.

Mallard E & Hugman S J, (1996) "The impact of mercury on the Bulolo river valley", Papua New Guinea University of Technology Research Report, 34 pp.

Nakanishi J, (1992) "Mercury pollution of Minamata, Canada and Amazon," *Water Report*, 2(4): 4.

Nakanishi J, Konuma S and Gamo M, (1994) "Risk evaluation of mercury pollution in the Amazon River system," *MIZU-JOHO (Water Report)*, 14(5): 12.

Nordberg G.F.(ed.), (1976) *Effects and dose-response relationships of toxic metals*. Amsterdam: Elsevier Scientific Pub. Co.

Subashinghe G.K.N.S, (1993) "Optimal Design of Sluice-boxes for fine gold recovery", *Minerals Engineering*, 6(11):1155-65, Pergamon Press.

Suzuki T, Imura N and Clarkson J W, (1991) *Advances in mercury toxicology*, Rochester Series on Environmental Toxicity, New York: Plenum Press.

3

The Bio-diversity Inventory: A Case Study of the Kamiali Wildlife Management Area

F.L. (Rick) Bein

ABSTRACT

A team of scientists during 1997-98 inventoried the biodiversity of the Kamiali Wild Life Management Area, 60 kilometres southeast along the Huon Gulf coast from Lae, Papua New Guinea. The 434 square kilometre area was registered with the Papua New Guinea government as a wildlife management area in 1996. A village of 500 people has elected a committee who with the Village Development Trust, a nationally based NGO, manage the area extending 12 kilometres out to sea and 17 kilometres inland. The high relief environment ranges from 1080 metres below sea level up to 2012 metres above mean sea level. A transect method was used to sample biomes beginning with open sea, coral reef, mangrove, beech, village, sago swamp, garden, riparian environments, tropical forest and cloud forest. This study provides a baseline inventory for a unique area which has never experienced hillside gardening nor commercial logging and contains a high biomass environment that serves as a habitat for many rare species. This case study demonstrates an application for the conservation of natural resources in rural Papua New Guinea.

The Role of Conservation and the Wildlife Management Area

"Conservation of resources" is a concept which lies half way on the continuum between "exploitation of resources" and "preservation of resources". Resources exploitation is the extraction of nature in a manner in which there is no regard for the long-term future benefits that those resources may provide for humans. The antithesis of exploitation is the preservation of resources which in the pure sense promotes the total non-use of the resources by humans. Neither of these extreme approaches can functionally serve human kind for the long term. Humans must use the earth's resources in order to survive and, conversely, humans will perish if they consume all of the resources at once. Conservation on the other hand promotes the sustainable use of natural resources.

Wildlife Management Areas (WMAs) have been established for years in a number of countries (including Papua New Guinea) to slow down the process of exploitation of natural resources. It falls slightly left of centre on the preservation side of the continuum. Another concept, the Integrated Conservation and Development Area (ICAD) is a recent addition to the many methods of protecting the natural environment from human destruction. The ICAD concept promotes sustainable human use of the resource and occupies a more central position. In Papua New Guinea, WMAs have been more commonly established because there are government regulations to follow. Regulations for the implementation of ICADs did not exist in Papua New Guinea until recently. In many situations the WMAs of Papua New Guinea are now including the thinking behind the ICADs by incorporating the concept of "sustainable use of resources." This is the current position of the Kamiali Wildlife Management Area (KWMA).

"Conservation" is quite different from "preservation" which promotes total disuse of the environment by humans. Although the idea of preservation has merit in small isolated parts of the earth, it is largely impracticable with growing populations and their increasing demands on the natural resources. Conservation on the other hand takes the position, "that if humans are to survive on this planet they must strive to find sustainable uses of the environment." Should people fail to do this they will perish. The planet has been around for a very long time and it will continue to be around for a very long time. The question is: "Will people continue to be along for the ride?"

Conservation of Biodiversity

Biological diversity is the richness of our planetary ecosystem and includes genetic diversity, species diversity and ecological diversity (Miller 1996). The conservation of the biodiversity is important not only to ensure survival of millions of species but also to support the living systems of the world's people who depend on natural resources and the ecosystem. Leading ecological thinking considers high biodiversity to be desirable for the planetary ecosystem. Following the ecological premise that every living thing has a "niche" or functional purpose in serving the greater ecosystem, it would hold that by maximising the number of living species, there would be more opportunities to maintain ecological balances. As a part of the global ecosystem, humans would benefit the most from high biodiversity. Benefits found in nature take the form of foods, medicine, building and clothing materials and chemical substances. Many discoveries have benefited human kind and have over the millennia fostered the expansion the earth's human population. Maintaining bio-diversity provides continuous human benefit. Preserving biodiversity does not contradict conservation. Conservation when operating properly preserves bio-diversity.

It further happens that tropical biomes such as those in Papua New Guinea contain some of the highest bio-diversity on the planet. Hundreds of species, many yet to be named, thrive in the Papua New Guinea rain forests and coral reefs. The world is eager to discover these new species, their ecological niches and the benefit they can provide human kind. Papua New Guinea is still 60 per cent forested, much of which is pristine and contains some of the richest biodiversity in the world (Filer and Sekhran 1998). There is great fear that Papua New Guinea will lose this resource if current trends of deforestation continue.

Papua New Guinea, on the other hand, is striving to develop economically and coupled with a 30 year population doubling rate it must exploit its basic resources. Traditional paradigms (or ways of thinking) and lifestyles of the village are changing. Attitude towards the environment is also changing. Western paradigms, promoting the monetary economy begin to erode the old lifestyles, and the environment becomes expendable for the purpose of seeking immediate wealth. During a paradigm shift the natural environment becomes vulnerable and easily exploited to support the new way of thinking. Exploitation occurs during the transition when old thinking gives way to, or merges with, new thinking.

The Kamiali Wildlife Management Area

Kamiali became a WMA because of several unique circumstances. Firstly, it is the only coastal village along the Huon Gulf between Lae and the border with Oro Province which chose not to sell their trees to the Malaysian logging companies during the early 1990s. Secondly, the food garden soils in the fertile Bitoi River flood plain provide enough food for villagers and there has been no need to clear the highly erodible hill sides. As a result of these two factors, much of the forest remains pristine, and the coral reefs fringing the coast of Kamiali are equally pristine because they are free from the smothering effect of sediments eroding from the land. Since Kamiali is mainly a fishing village, there is very little hunting pressure and the original wildlife is intact. The seven village hunters have had little problem finding wild game within short distances of the village and as a result rarely venture up into the more remote areas generally above 300 metres elevation.

The Biodiversity Survey

The biodiversity inventory was conducted between July 1997 and July 1998. When the Lababia community established the Kamiali Wildlife Management Area, the Village Development Trust of Lae, Papua New Guinea, obtained a grant from the New Zealand High Commission to conduct this study. The grant provided for the listing of species found in the various biomes throughout the Wildlife Management Area and the listing of recommendations regarding further study and better conservation of the wildlife management area. The sampling technique comprised a transect from the open sea through the coral reef, mangrove swamp, beach, freshwater mangrove, sago swamp, lowland rainforest, lowland hill forest, upland and montane cloud and moss forest. Also sampled were sites that were deemed to be notably different such as Lababia Island and the Bitoi food gardens. No attempt was made to numerically quantify any part of biodiversity except in the case of the fish study. Nor was there any attempt to sample the genetic diversity.

A signed "scientist, good faith, research agreement" has been established to provide an incentive for the villagers to host scientific researchers and to protect their intellectual property rights. This includes an understanding that participating scientists

- build into their research proposals and grants a cash contribution to the community,
- employ villagers when ever possible in the data collection and educate them within their desire to understand about the research and its methodology in order to strengthen their abilities and enthusiasm to support future research, and
- commit to share the proceeds, should the scientist or representative institution make a profit from their discoveries at Kamiali, as a continuing contribution designated for worthwhile village causes such as the payment of school fees or for the purchase of environmental conservation technology.

Facilities to accommodate field research scientists include the comforts of a guest house with prepared meals, and a training centre built by the Village Development Trust, air compressor to refill diving tanks, storage capability, easy access to most biome areas, transect trails to the more remote environments and boat transportation from Lae. Up to 16 people can be accommodated in four-person insect screened bunk houses and another 20 people in a dormitory bunk room. Running water supplies clean showers and toilets.

Preliminary Results

Thirteen participating scientists focusing on different parts of the biota have found many unique aspects of the biodiversity of the Kamiali area. Briefly, the highlights of the findings are as follows:

- The fringing coral reef is one of the highest biomass reefs in Papua New Guinea.
- The undisturbed lowland hill forests next to the ocean shoreline are unique because they have not been periodically cleared for food gardening like most Papua New Guinea coastal communities.
- Vegetation typical of 2000 metres elevation is found as low as 600 metres on ridge slopes facing toward the sea.
- A vast area of land has not been touched by humans for at least the last 50 years.

- Lababia Island serves as a major roosting site for frigate birds and pied imperial pigeons.
- The broad sand beaches of Kamiali are major nesting sites for leather back turtles.
- Several undescribed insect species are in the process of being recorded.

The complete inventory contains the appended reports of each scientist including all species found and those species, not seen, but which are probably there because of their association with the known species. Also included are recommendations made by the scientists for maintaining the biodiversity and for future research.

A Base for Further Research

The biodiversity inventory at Kamiali with its logistical facilities in place provides an excellent basis for further research. The potential for future research in such a pristine environment is unending. The major difficulty is the long distance travel and time required to come to Kamiali from the research institutions around the world. This inventory is not all-inclusive as the sampling technique used could only probe a small area. There are great research possibilities in the further sampling of the biomes. The cloud forest biome has been barely examined for plants and insects. The coral reefs and mangroves and beech areas offer many opportunities. Over 100 leather back turtles nested in Kamiali this last season and discussions are underway to experiment by incubating some of the eggs to ensure greater hatch rates. Ethno-botanical studies were only started with the inventory while an entire research agenda could be developed to study the medicinal plants and herbs.

REFERENCES

Filer,C. and Sekhran, N. (1998) *Loggers, Donors and Resource Owners*. Port Moresby: National Research Institute.

Miller, G. T. Jr. (1996) *Living in the Environment* (9th ed). Belmont, Calif: Wadsworth Publishing.

4

Logging in Papua New Guinea: Policy Formation, Implementation and Forest Sustainability

Philip Siaguru

ABSTRACT

There is no other government entity that has received so much attention in the last five years as forestry. Attention to forestry began with small logging extractions for buildings during the colonial era. This attention accelerated after the war years as increased local timber consumption for reconstruction purposes developed into current commercial logging. These developments had their consequences on the economy and the environment of Papua New Guinea. The situation changed dramatically after the Barnett Inquiry revealed massive corruption in the forestry sector in 1989. A new Forest Policy of 1991 and a new Forestry Act of 1991 (Amended 1993) changed the forestry sector. Structural reforms began at Hohola in the Department of Forests and spread through all the appropriate sections of the government, people's organisations and the forest industry in provinces and the villages. Several initiatives were undertaken by the Government of Papua New Guinea with international assistance. Landowner groups were consequently incorporated. This initiative is to encourage the villages in the process of tapping their forest resources, making them subjects and not objects in development. These initiatives are discussed here in relation to the Forest Policy and Act. Some of the more salient points are highlighted. It is hoped that this information will provide the basis for more regional collaboration in the process of achieving sustainable management of logging not only in Papua

Cont'd...

New Guinea, but also in the region and the world as a whole. The different initiatives generated by concerned managers are presented purely as terms of reference. It will take time to see the effects of these initiatives, since planning has it that the fruits of these different investigations and written documents will culminate by the year 2000 and beyond. Hence, it is hoped that items, factors and views presented in this paper can be a forum for continued discussion on the subject of forestry.

Forests and Forestry in Papua New Guinea

The National Goals and Directive Principles complement to a very large extent the fourth goal of the country's Constitution which relates to the Natural Resources and Environment: ". . . for Papua New Guinea's natural resources and environment to be conserved and used for the collective benefit of all, and be replenished for the benefit of future generations."

Papua New Guinea's forests cover a total area of 36 million hectares. There are over 9000 species of higher plants, including well over 400 species of forest trees. These forests support and maintain about 700 species of birds including the world's smallest, some of the largest parrots and pigeons, all of the world's three species of cassowaries, and more than two-thirds of the known birds of paradise (United Nations, 1994). Papua New Guinea's forest is also home to the world's largest butterfly species, two hundred species of mammals which include two of the world's three monotremes and all of the world's tree kangaroos (United Nations, 1994). Above all, these forests also support and maintain the rural population of about 3.7 million (1993 census). Many people have mentioned that the forests are probably more valuable for their biodiversity than for their logging potential. Approximately 15 million hectares of Papua New Guinea's forest are accessible and operable for logging. Of the 15 million hectares 1 million hectares, mostly lowland forests, are already harvested. The state has acquired an additional 5.8 million hectares for logging (Papua New Guinea Forest Authority, 1995a).

The primary responsibility of any land manager in Papua New Guinea is the protection of the biodiversity in ways that allow replenishment. The responsibility is for resource owners to benefit from the protection of their forests. Logging is one source of income for the villagers. The United Nations has reported some initiatives currently in progress in Papua New Guinea. The Global Environmental Facility (GEF) provided an initial grant of US$5 million to establish a Conservation Resource Centre and two pilot Integrated Conservation and Development (ICAD) projects. The Chevron Kutubu oil project has also established an ICAD project in Gulf province.

The current situation in the forestry sector in Papua New Guinea is that out of a total land area of 66.9 million hectares 36.4 million hectares are under forest cover. Accessible and operable forest is 15 million hectares, 5.8 million hectares of which has been acquired by the state. The state further acquired and has allocated 4.4 million hectares of loggable forests. There are 2.4 thousand hectares of national forest areas and a further 63.9 thousand hectares are earmarked for forestry development. Another 37.4 thousand hectares are granted as plantation areas. The sustainable harvest volume from the available natural forest has been estimated at 3-4 million cubic metres per annum. The current permitted cut has been placed at 7.8 million cubic metres, but the actual cut has remained low. From 1 million cubic metres in 1980 it has risen to an estimated cut of 3.5 million cubic metres in 1994.

The Barnett Inquiry into the Papua New Guinea forestry sector was concluded in 1989. The inquiry revealed substantial malpractices. Since then the government has responded to the Barnett Inquiry in a number of ways. Its first action was the passing of a new *Forestry Act* in 1991 (Amended 1993), which created a new Forest Authority (NFA). The NFA is governed by a Board which has representations from the private sector, NGOs and Provincial and National Government representatives. The government also initiated and steered into establishment the National Forestry and Conservation Action Programme (NFCAP) with four main tasks. These tasks were to increase landowner awareness of the economic and non-economic values of their forests, to promote landowner involvement in planning, managing, conservation and the utilisation of forests, and understanding and protecting forestry diversity in Papua New Guinea to ensure long-term employment and sustained economic returns from forest utilisation for the benefit of the nation and the resource owners.

The 1991 *Forestry Act* (Amended 1993) emphasised sound forestry practices on a sustained yield basis. This requires that the logging activities of companies, both their logging practices and exports, are to be monitored adequately. Regeneration is also encouraged through improved silvicultural management by individual export companies and resource owners. The *Forestry Act* empowers the Forest Policy which was formulated after collaboration between the National Government, the industry and the Provincial Government. The main objectives of the Policy are to ensure that the sustainability of forests and the renewability of forests is maintained through sound management principles and that harvesting should promote economic growth through job creation by onshore domestic processing of logs and thus encourage increased participation by Papua New Guineans. This policy was adopted in September 1991 by the National Executive Council.

Further development in the regulation of forestry practices eventuated in the production of the National Forestry Guidelines of 1993. Section 47 of the *Forestry Act* specifically calls for the national forest plan to be consistent with the National Forest Policy. This in fact became the marriage clause which provided for sustained yield harvest and management of Papua New Guinea's forest resources.

The Historical Background to Current Forest Policy

The last five years (1991 - 1995) have seen substantial changes in the forest policy. The situation and circumstances leading up to the commissioning of the Barnett Inquiry and the looseness of the policy before 1991 are of importance. The first central government in Papua New Guinea was constituted in Papua in 1884. Forestry was not included. The administration placed importance on political governance. In 1908, a forestry officer from Queensland (Australia) assessed the timber potential of Papua New Guinea. Nothing was done after that assessment. In 1921 a fuller report assessing the natural resources was undertaken and a report was prepared by Lane-Poole in 1924 which recommended a policy and ordinance, but no action was taken. It was not until 1948 that McAdam (the first Director of Forests) got proposals approved for a forest resource survey, a programme of forest reservation, a programme for research, management and working plans, and training of staff. In 1951, the Minister for External Territories made a forest policy statement to this effect: (i) to locate, assess and regulate availability of natural forest resources to bring them within reach of development, (ii) to encourage investment of private capital for development of these resources, and (iii) to ensure native participation in this development.

The emphasis of the forest policy was on sawn timber for post-war reconstruction and land clearance for agriculture. In 1957, a further policy statement approved the acquisition and reservation of an adequate permanent forest estate of 4 million acres (1.6 million hectares) of dedicated forest within 10 years and 10 million acres in 20 years. It also approved the establishment of a forest training centre and a forest research institute. A re-afforestation programme including expenditure in the Highlands for grassland reclamation was also established. The policy recommended a combination of research and development through timber utilisation research into natural durability and preservative treatments, botanical collections and identifications, and encouragement of a timber industry.

This policy statement and its subsequent implementation provided the basis on which development has taken place. The above policy statement guided forestry in Papua New Guinea until 1973, when a policy statement by G.L. Garson announced seven major objectives based on two factors. First, it acknowledged the people that existed and dwelt on the land and, second, it attempted to translate the government's improvement plan into a National Forest Policy. There were seven major objectives. The first objective was for Papua New Guinea's forest resource to be managed as a national asset in the interest of the present and future generations of people (who are forest resource owners). The second objective was for Papua New Guinea to preserve, develop and maintain through reforestation such forest areas as will enable domestic needs to be met as far as is possible and is warranted and enable full advantage to be taken of export opportunities. The third objective was for Papua New Guinea to accept the need for protection and management of watersheds, control of soil erosion, conservation of animal and plant communities and the use of forest for recreational and other direct benefits. The fourth objective was to encourage the development of permanent forest industries and to provide opportunities for the forest owners to become involved. The fifth objective promoted research into forest technology in order to improve the efficiency of forestry and forest industries. The aim was also to reduce imports and increase export earnings leading to an improvement in the social and economic life of the nation and the forest owners. The sixth objective promoted education in all branches of the forest industry and to encourage a better understanding by the people of the value of their forests in both their own and the national interest. The seventh objective was to provide the laws, the organisation and funds to manage effectively the nation's forests having regard to overall development policies and the forestry sector's ability to compete successfully with other demands on the nation's resources.

In 1974, the then Minister of Natural Resources, Mr. Bruce Jephcott, published a forest policy statement in which an attempt was made to translate the National Eight Point Improvement Plan into forest policy terms. The statement promoted a rapid increase in the proportion of the economy under the control of Papua New Guineans and a move towards the equal distribution of economic benefits. The decentralisation of economic activity planning and government spending with an emphasis on agricultural development and village industry came third on the list. Emphasis was to be placed on small scale artisan services and business activities. The aim was to promote a more self-reliant economy less dependent for its needs on imported goods and services. The contribution was an increase in capacity for meeting government spending needs from locally raised revenue. A rapid increase in the active participation of women in both economic and social activity was as important as government control and involvement in those sectors of the industry where control is necessary, to achieve desired kinds of development.

Another document on the forest policy was the White Paper prepared and approved by the National Executive Council on 23rd May 1979. That statement rightly starts by considering the National Goals and Directive Principles as contained in the Constitution of Papua New Guinea and the eight aims in the national development strategy. It focuses on issues relating to the forestry sector. It also contributed meaningfully to developing national objectives with particular scope to generate revenue, greater national ownership with regional economic development and political stability. This was a shift from the policy prior to 1979 which stressed industrialisation and local processing (phasing out of log exports) to be achieved primarily through foreign investment.

The forest policy statements before 1973 totally ignored people's participation and never even attempted to define land and resource ownership according to the Papua New Guinean concepts. That was the major flaw in that era of policy development which still has not changed even to this time. No one has yet defined a new radical policy that will bring more incentive to the resource owner. There is no political or social will to do so. The 1979 policy statement retained all the seven points stated in the 1974 statements, emphasising the principles of local processing. A number of nationally-owned forestry companies emerged. This eased restrictions on log exports. Export tax was also reviewed and increased. One objective of the 1979 policy was met with resounding success. Export figures presented by the Department of Forest (1990) (See Table 4.1) showed an increase of log exports almost 3 times from 1979 to 1988.

Efforts to implement other aspects of the policy were eventually forgotten. The number of logging operators increased since export of logs was the main line industry. National ownership was not encouraged. The processing capacity declined and deteriorated and almost all foreign investment went into logging and exporting ventures with no concern at all about local capacity building and improvement of technology. Almost all efforts of the 1979 policy were based on encouraging large log export quotas for companies in exchange for the building of infrastructure and establishment of agricultural projects. All powers were tied up in the hands of the Minister and the central government. No efforts were made to devolve power to the provinces.

Table 4.1 Figures below show exports of forest products (000 m3 and millions in Kina)

	1979	1980	1981	1982	1983	1984	1985	1986	1987	1988
Logs	472	642	743	1063	1019	1284	1158	1314	1442	1361
Sawn Timber	63	45	24	21	20	18	15	7	4	3
Plywood	6	6	8	6	6	5	1	2	0.02	
Woodchips	107	121	103	148	100	103	82	61	66	67
Value	36	49	45	53	56	82	67	77	113	100

Source: Department of Forests (1990)

Recent Changes in Forest Policy

An initiative taken by the government of Papua New Guinea as a consequence of the Barnett findings (1987-1989) was to review the whole forestry sector. On Papua New Guinea's invitation the World Bank led the review with smaller donor agencies assisting in different smaller projects. The report recommended the formulation of a new policy, legislation and a programme of technical and financial support. After the report was accepted by a meeting between the Papua New Guinea Government, local and international NGOs and donor representatives in 1990, the Papua New Guinea Government began to reform the Forestry sector through the National Forest Conservation Action Programme (NFCAP). The reform was geared to three overall aims: (i) increasing landowner awareness of the economic and non-economic values of their forest and promoting their involvement in planning and managing the utilisation and conservation of forests; (ii) understanding and protecting forest diversity in Papua New Guinea ; and (iii) ensuring long-term employment and economic returns from forest utilisation, for the benefit of resource owners and the country as a whole.

Phase one of the programme began about 1990 and ended in 1994. The 1985 Forestry Sector Review Mission identified 14 areas that needed reformation. The Department of Forests and Department of Environment and Conservation presented a list of priority projects to donors. These include the World Bank as the leading donor agency, together with other agencies including the Asian Development Bank (ADB), The United Nations Food and Agriculture Organisation (FAO), the United Nations Development Programme (UNDP), the International Institute of Environment and Development (IIED), the International Union of Conservation and Nature (IUCN) and the bilateral development assistance from Australia, New Zealand and Germany. The priority areas, under the Department of Forests leadership, were identified as rapid resource appraisal, forest resource assessment, landowner awareness and support, improved ecological survey and monitoring, and human resource development. The Forest Management and Monitoring Task Force, strengthening the State Purchase Option, forest industry development studies, technical support for TFAP implementation, forest revenue study and corporate planning, agroforestry, forest resource information, mapping and database, and ecological and economic sustainable tropical forest use also came under the same category. The Department of Environment and Conservation (DEC) took responsibilities in conservation and world heritage areas, rehabilitation of existing National Parks, strengthening environment and conservation management, Moitaka offices, and provided immediate technical support. The Wau Ecology Institute and landowner awareness and support programme came under the NGO leadership.

Phase one of the NFCAP (National Forest Conservation Action Programme) achieved partial success. The following are initiatives that are known to have been achieved. The first of the successes was that the National Executive Council approved the National Forest Development Guidelines. In 1991 the National Forest Act was gazetted after the formulation and revision of the national forest policy. The *Forestry Act* of 1991 was designed to ensure the sustainable use and protection of the forest for the benefit of the indigenous landowners and the country as a whole. The *Forestry Act* was amended in 1993 with the main change being the reduction of the Forestry Board membership from eleven to six. By this time section 143 was also included in the Forestry Act. The original eleven-man Board consisted of the Departmental Heads of Environment and Conservation, Trade and Industry, Finance and Forestry; one each from the four regions in Papua New Guinea (Northern, Southern, Islands and Highlands), one appointed by the Minister of Forests with appropriate experience in commerce and finance, and an NGO representative. With the reduction, the four regional representatives were excluded together with the membership of a person versed in commerce and finance. The inclusion of section 143 in the amended Act was to authorise the Forestry Board to extend any logging permits permitted under section 137. Section 137, in turn, was originally included to honour logging permits which were issued under the old logging agreements of Timber Rights Purchase, Private Dealings Act and Native Timber Authority. The legal establishment of the National Forest Authority (NFA) was designed to enforce the new legislation with representation from provincial governments, landowners, NGOs and the private sector, but the operations of the National Forest Authority were heavily constrained by a chronic lack of funds. The new legistation covered Conservation Needs Assessment (CNA). Maps and database of Papua New Guinea's biodiversity were included. The Department of Environment and Conservation approved the strategic plan and an outline of the National Sustainable Development Strategy plan. International investment in integrated conservation and development (ICAD) programmes, and in forest sector projects, and the formulation of a national alliance of NGOs (NANGO) were also included.

There has been a complete structural change in the forestry sector. On October 8, 1993, the Department of Forests became the new National Forest Service (NFS). The new National Forest Service is now headed by a Managing Director. This new management team has one goal, that is, to manage the forest resources of Papua New Guinea as a renewable natural asset for the well-being of the present and the future generations. The National Forest Service is managed through a central headquarters based in Port Moresby, whose chief role is to set policy and provide specialist services for the four regional offices (Highlands, Islands, Northern and Southern). Each regional office serves provincial management committees and undertakes project level planning and supervision of field staff in the different provinces. NFCAP was embarked on as a very ambitious programme since it involved several government departments, international development agencies and NGOs (international and national). The legal bond holding all these is the *Forestry Act* passed in 1991 and amended in 1993.

Efforts to strengthen conservation management were begun by strengthening the staff of the Department of Environment and Conservation and developing a new project focus. These efforts began in November 1994. So far the staff and capacity strengthening is progressing well. The broad scale biodiversity mapping for Papua New Guinea has been completed and is being used in formulation of provincial forest plans. Mobilisation of Global Environment Facility (GEF) identified two major areas and funding was made available. This initiative was aimed at developing resource owners' interests and balancing that with conservation plans. At this present time, no information can be obtained from authorities on the progress of the GEF to date.

Research priorities in Papua New Guinea have recently come under the microscope. This was essential, since after the *Forestry Act* and the Policy were passed, new dimensions in forestry operations were contemplated. The aim is to provide a scientific basis for the management of Papua New Guinea's forests.

The Forest Research Institute has refocussed its operations with emphasis for sustainable management. Its current task is to identify species attributes and growth information in logged over forest. This is crucial to predicting growth models for determining future yields and cutting cycles. Research also supports alternative silvicultural options. This is essential, since there is none currently being applied to natural forest harvesting and replacement of the forest. The limited plantation research being undertaken is aimed at improving the productivity of industrial plantations and providing the information needed for community based woodlot plantings in areas where fuelwood and local building materials are in short supply. Moreover, given the very limited funding available for research, the refocussing of priority activities has meant abandoning several past initiatives and retrenchment of staff. This has trimmed the Forest Research Institute to basic manpower level. To run the Institute effectively and maintain good research development would require about six times the budget currently available, which is about 1.5 million Kina.

The *Forestry Act* of 1991, amended in 1993, is intended to achieve its goals in a number of ways. Section 46 of the Act states that the rights of the customary owners of the forest resource shall be fully recognised and respected in all transactions affecting the resource. Hence, this section unequivocally accepts the fact that any exploitation of forest resources will only be allowed with the consent and participation of the traditional landowners. Section 47 deals with National Forest Plans. It enumerates the steps that are to be taken to ensure sustainable development. It calls for development that must conform with National and Provincial planned programmes. It must take note of the set allowable harvest cut within the framework of sustainable management. Section 49 specifically commits each Provincial Government to consult closely with the Provincial Forest Management Committee (PFMC) so that any plan that is developed will conform to the National Forest Development Guidelines.

Resource acquisition follows only after the PFMC has perused the project guidelines and recommended the necessary appropriate guidelines to the Forest Board for approval. (See Figure 4.1) Thus, the PFMC is involved right at the beginning of the process. It is only after this process that the project is advertised and intending bidders may make project proposals which are screened initially by the PFMC. After this step, if the Minister accepts the proposal, he will require approval of an environmental plan of the area by the Department of Environment and Conservation. If the environmental plan is approved the permit is granted. The permit holder is expected to provide a project statement outlining plans and objectives, a five year working plan and an annual logging plan. The Forest Officers continue to inspect and conduct checks to ensure that the annual logging plan has been followed. These inspections also involve checking of documents kept by the permit holder which often contains specific species by species harvest schedules.

Hence, a comprehensive and serious review has been done after the new Forest Policy and Act of 1991. The forestry sector has received

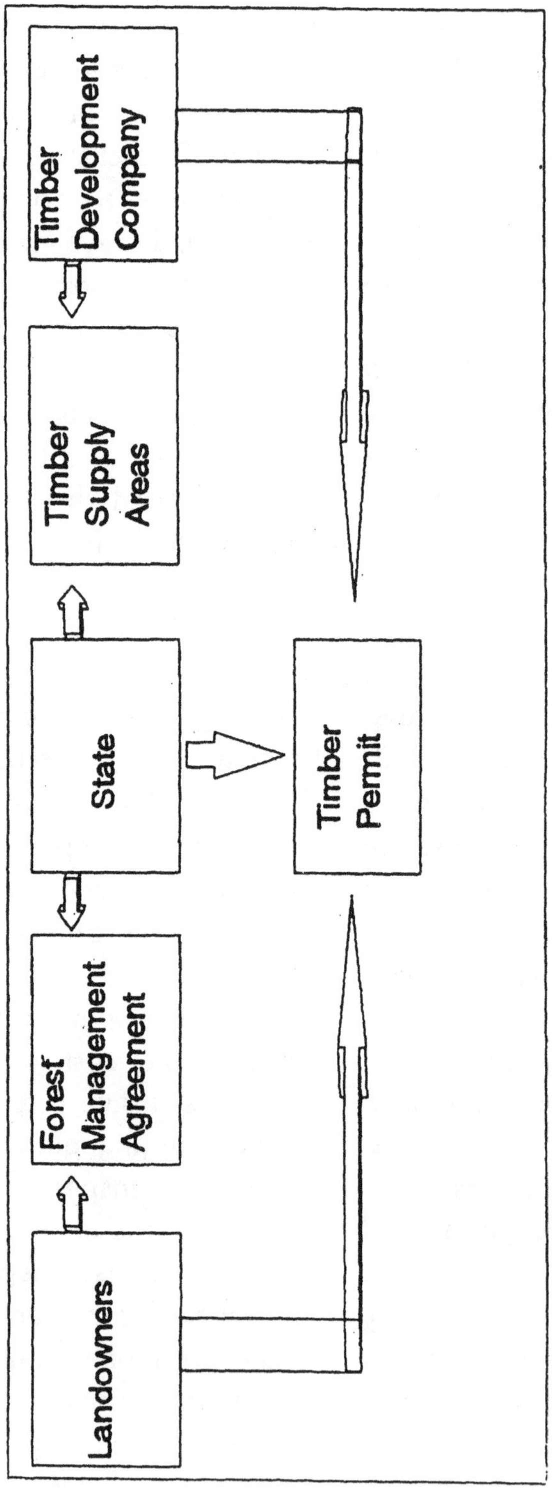

Figure 4.1 The procedure that is now being followed to acquire forest resources

in the last five years more reviews and comments by consultants and experts than any other government department in the Papua New Guinea public sector. These efforts have resulted in the production of several documents to assist planning and management. These are the National Forest Policy of 1991, the Forestry Amendment Act of 1993, the National Forestry Development Guidelines, the Guidelines for Issuing Timber Authorities, Guidelines for the Issuing of Provincial Plans, Papua New Guinea Logging Code of Practice, Instructions for Log Exporters - Interim Changes in Procedures, and Key Standards for Selection Logging in Papua New Guinea, Directions for the identification, scaling and reporting on logs harvested from natural forest logging operations and Manual on land group incorporation are part of the amended Act of 1993.

Two other National Government agencies, Department of Finance (DOF) and the Department of Environment and Conservation (DEC) are responsible for the legislation which has a considerable influence on the forest harvesting. The DOF enforces the *Customs Act*, *Customs Tariff Act* and *Exports (Control and Valuation) Act*. These Acts control and regulate the import and/or export of forest produce. The DEC controls conservation and environmental issues. The Acts supporting this are the *Environmental Planning Act*, *Conservation Act*, *Water Resources Act* and *Environmental Contaminants Act*. The *Environmental Planning Act* provides legislation where the environmental implications and effects (physical, biological and social) that have direct influence resulting from development is assessed. To a very large extent this complements the *Forestry Act* which states that a timber permit allowing harvesting cannot be granted without an approval under the *Environmental Planning Act*. This measure is applied wholly and requires the preparation and submission of an Environmental Plan to specifications provided by DEC. The *Conservation Areas Act* allows for the establishment and management of a particular area for conservation purposes. The *Water Resource Act* governs the management and use of the natural waters and is administered through the use of a permit system and the *Environmental Contaminants Act* contains a regulatory framework for the control of environmental contamination and the licensing of importing, using and disposing of hazardous substances.

Contemporary Problems and Regional Initiatives

In a paper presented at the third *Obihiro Asia and the Pacific Seminar on Education for Rural Development* (Siaguru 1994), I portrayed an eco-unit consisting of five main factors. These are (i) village displacement, (ii) flora and fauna extinction, (iii) imbalance of village existence, (iv) reduction of economic potential, and (v) environmental degradation. All these five factors must be taken as a whole unit of sustainable management. Preserving all five areas and maintaining them was what I referred to as sustainable management. But these responsibilities lie in the hands of the local people of Papua New Guinea. The people must have the will to implement and carry through all the suggestions that have been provided through all the different government initiatives such as the *Forestry Act*, the Forestry Policy and all the other initiatives taken by the government. Only when that happens can sustainable management be achieved.

It is crucial and we all must understand that the villager who lives and dwells in the forest should be offering his 'supermarket' for logging only after careful consideration of the opportunity cost involved. Logging is seen by many rural landowners as their best means of moving into a cash economy and in obtaining facilities such as schools, roads, bridges and medical services. The political pressures on the Forest Authority are enormous. While that is true on the one hand, on the other, there are vocal and radical landowner groups demanding equal attention to the protection of forest. This alternative is gaining in popularity now that some NGOs have begun to promote programmes that assist villagers develop alternative means of generating a cash income.

A great deal of progress has been achieved in the last five years in Papua New Guinea. It is hoped that this will provide a solid foundation on which sustainable management of the forest resource can be achieved. It must be appreciated that in order to completely undo past logging arrangements, the Papua New Guinea Forest Authority has compromised some forest areas for harvesting at unsustainable levels until the current permit expires. The government of Papua New Guinea through the Papua New Guinea Forest Authority hopes that the support given it through international assistance is maintained until Papua New Guinea is able to demonstrate to the world that it is irrevocably committed to sustainable forest management.

On a regional scale the Melanesian Spearhead Group (Papua New Guinea, Solomon Islands and Vanuatu), experience similar pressures on their forest resources and could benefit by sharing their experiences. Hence, it has been agreed in its summit meeting in July 1994 that the agencies involved will cooperate in the following areas: intelligence gathering regarding the interconnecting relationships of foreign investors in the timber trade; market price and volume data exchange; development of inter-country cooperative marketing of production from round logs; understanding the methods utilised to achieve price transferring and other illegal activities; forest operational techniques for standardisation of forest monitoring and control; improvement in the level of forestry expertise available to the Melanesian people through education and training; and exchange of scientific information. Also at the 25th meeting of the South Pacific Forum in 1994, the Prime Ministers of Australia, Papua New Guinea, Fiji, New Zealand, Solomon Islands and Vanuatu agreed that they would work towards a common code of conduct governing logging of forest resources. Under the common code the logging companies would face similar operating conditions in different countries. There is also a need to urgently increase the monitoring of logging and the export of timber. In Papua New Guinea these resolutions have resulted in the involvement of four forest rangers from Australia currently working alongside the Papua New Guinea Forest Authority to assist in the implementation of the code of logging practice.

Conclusion

A review is currently being made of the Constitution of Papua New Guinea. Natural resources have been included in the terms of reference to be discussed for possible expansion. It is argued that, except for the fourth goal which calls for the use of natural resources for the collective benefit of its people and for them to be replenished, nowhere else in the constitution are the forests protected. Hence, the review hopes to expand and make preservation of natural resources more a part of the nation and the country by specifically including it in the Constitution.

While the philosophy sounds good, the reality and its practicality remain very doubtful. Figure 4.1 shows the new procedure to obtain forest resources. While it appears simple and straightforward on paper, in reality it is not. Foreign investors are directly seducing illiterate and unsuspecting villagers with cash to win favour and gain their approval for logging. This is done directly, hence, the process of approval by the Board becomes a mere formality, since villagers' concerns can still influence the PFMC recommendation to the Board of the National Forest Service.

Hence, in my view, the issue is not to change and review the Constitution but rather to improve on monitoring and the supervision of logging deals. Such policing must involve all the government departments responsible for managing the land. Isolated individual operations by the different government arms such as the Department of Environment and Conservation, Department of Forests and Papua New Guinea Forest Authority do not get Papua New Guinea nor its people anywhere. Views from both parties (Government and people) are important in the sustainable management of the forest resource. By the same token, both parties must be ready to compromise, develop and accept the best model to take Papua New Guinea into the year 2000. International NGOs (including World Bank) must recognise these concerns and assist appropriately and not merely try to influence and direct local capacity building based on assumptions and experience gained from other tropical countries.

REFERENCES

Department of Forests (1990). Monthly Export Statistics. Port Moresby: Government Printer.

Government of Papua New Guinea (1975). *The Constitution of Papua New Guinea*. Port Moresby: Government Printer.

The Asia Pacific Action Group (1990). *The Barnett Report. A Summary of the Report of the Commission of Inquiry into Aspects of the Timber Industry in Papua New Guinea*. Hobart: APAG.

Papua New Guinea Forest Authority

(1991a). *National Forest Policy*. Ministry of Forests.

(1991b). *National Forestry Act*. Ministry of Forests.

(1995a). "The Papua New Guinea Forestry Sector. Briefing notes for the incoming Managing Director of Forests."

(1995b). "Papua New Guinea Country Statement. 1995 mid-term review of progress towards the achievement of the ITTO year 2000 objective."

Siaguru, P.(1994) "Utilisation of Papua New Guinea Forests in Harmony with their Conservation," n *Utilisation of Forests in Harmony with their Conservation: Proceedings of the 1994 Obihiro Asia and the Pacific Seminar on Education for Rural Development*. Obihiro University of Agriculture and Veterinary Medicine, Hokkaido. Japan.

United Nations (1994) *Yumi Wankain*. Report of the United Nations Joint Inter-Agency Mission to Papua New Guinea on Sustainable Development.

5

A Ram Pump Primer

Brian Young

ABSTRACT

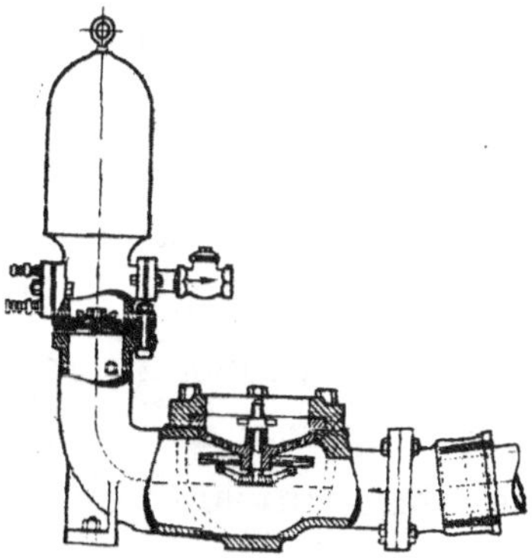

The hydraulic ram pump was invented over 200 years ago but became obsolete with the availability of mains water. Ram pumps make use of a renewable energy resource and are therefore environmentally friendly. Renewable resources such as wind, solar energy and falling water also represent economic and environmentally friendly sources of power. Whilst wind and solar energy tend to be used for pumping groundwater, pumping of surface water using the potential energy of a plentiful supply of the same water is possible with a ram pump. These pumps are particularly appropriate for use in the developing world as their construction is basic and robust and they are inexpensive and easy to install and maintain. The purpose of this paper is to de-mystify the ram pump and to encourage their use for rural community water supply in Papua New Guinea.

Introduction

The automatic hydraulic ram pump was invented more than 200 years ago. It requires no external power source but makes use of the potential energy of the water itself to pump a small proportion of the supply to a considerable height. It is environmentally friendly, has only two moving parts and may be purchased relatively inexpensively (depending on size) or be home-made from suitably modified standard pipe fittings.

Hydraulic rams were used mainly in Europe, Australia and the United States during the nineteenth and early twentieth centuries for the supply of water to farms and country estates but, with the introduction of piped water and electrical supplies, they became obsolete. Recent enthusiasm for low cost, low maintenance, environment friendly devices using a renewable energy resource has rekindled interest in their use for water supply in remote regions not served by public utilities.

Other pumping systems, employing renewable energy directly from the wind and the sun, may be used for pumping ground water. The ram pump, however, pumps surface water and requires a fall of the same water to make it work. A fast flowing mountain stream or small waterfall providing a plentiful supply of clean water is ideal. Typically, a pump would be sited a few metres below the intake and is capable of pumping a small percentage of the supply to heights over 100m. Although not discussed here, the compound ram pump is capable of using the power of a polluted stream to pump a separate supply of clean water. Pumps may also be operated in parallel (each having their own drive pipe and supplying a single delivery pipe through a manifold) but this option tends to be more expensive than a single pump of equivalent size.

Very little in the way of environmental impact occurs with the manufacture or installation of ram pumps. The construction presently favoured in Papua New Guinea involves the assembly of imported pipe fittings and minor machining for the production of the impulse and delivery valves. Installation requires only the construction of a small dam and foundations for the pump unit itself: delivery and reticulation pipelines are buried and no heavy machinery is required. Nature quickly takes over a site and there are even instances of ram pumps being "lost" to succeeding generations.

How a Ram Pump works

A ram pump system is more than just the pump unit itself. Figure 5.1 shows the complete

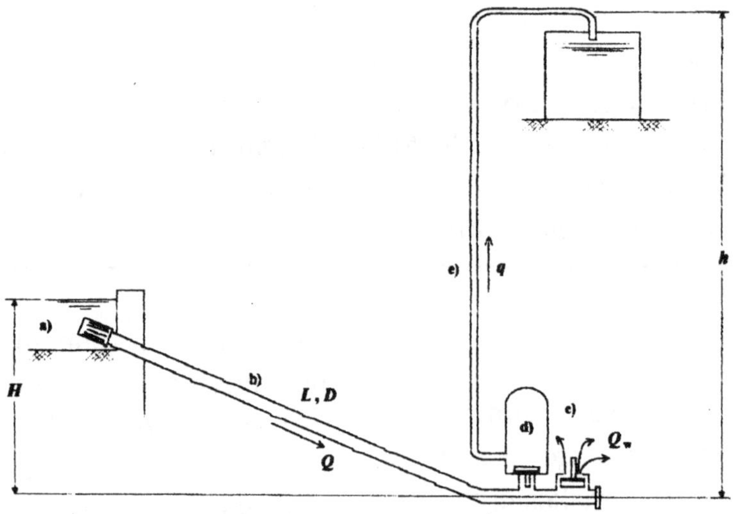

Figure 5.1 Layout of a ram pump system

system consisting of: a) intake at a head H above the pump unit, b) drive pipe of length L and diameter D, c) pump unit with impulse and delivery valves, d) air chamber and e) delivery pipe to the storage tank at a height h above the pump unit.

The function of the air chamber is to store water at the delivery head between strokes and to absorb the shock of pumping. The air chamber must always be charged with air and this is sometimes achieved by the inclusion of a small "snifter" valve placed near the delivery valve.

The operating cycle of an automatic hydraulic ram consists of three distinct phases (see Figure 5.2): i) acceleration, ii) pumping and iii) recoil. Acceleration of the water in the drive pipe occurs when the impulse valve is open and the delivery valve is closed. If the valve were to be held open, the water velocity would eventually reach the steady-state value, u_o. However, under normal conditions and at a certain critical velocity, u_c (which depends on the impulse valve geometry, stroke and weight) the valve closes abruptly due to hydrodynamic forces overcoming the dead weight or spring load tending to keep the valve open. Pumping now takes place as pressure pulses (water hammer) induced by the sudden stoppage of water, pass up and down the drive pipe at the speed of sound, the delivery valve opening against the back pressure of the delivery head in response to each pressure pulse. Recoil, or the reversal of flow in the drive pipe, can occur at the end of the pumping phase after closure of the delivery valve. The suction resulting from the recoil (or the effect of dead weight or spring load in the absence of recoil) causes the impulse valve to open and the cycle is ready to begin again.

Beat frequencies for ram pumps can be low (30 per min) or high (120 per min) depending on values of the design parameters for the system. Increase in impulse valve stroke or weight and drive pipe length decreases beat frequency; increase in supply head increases beat frequency.

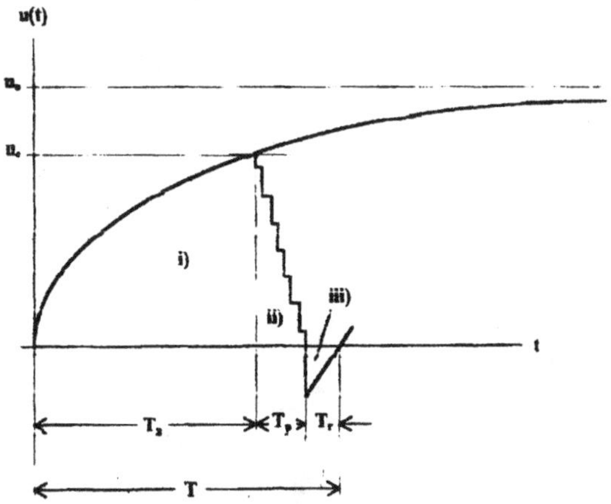

Figure 5.2 Drive pipe velocity-time diagram

The maximum delivery pressure, h_m of which the pump is capable is directly proportional to the velocity of sound in the drive pipe and to the maximum or critical velocity u_c reached by the water in the drive pipe during the acceleration phase. Clearly, if the delivery head h is equal to, or greater than, h_m no water will be pumped. The magnitude of the critical velocity depends on the design of the impulse valve but for a particular ram, u_c can be increased by increasing the valve mass and stroke. Obviously, the critical velocity cannot be increased indefinitely since, if it reaches the steady-state value, the valve will not close. Ram pumps should normally be operated at less than 60 per cent of h_m since performance deteriorates rapidly above this value.

The velocity of sound in the drive pipe is determined by a number of factors such as drive pipe material and wall thickness and the volume fraction of air trapped in the pipe. For a steel drive pipe of standard wall thickness with no entrained air, the acoustic velocity is about 1350 m/s. Air can be sucked into the drive pipe during the recoil phase and this has a marked effect on the acoustic velocity. For example, one-tenth of one per cent by volume of undissolved air in the water will cause a fifty per cent reduction in the acoustic velocity. The amount of air sucked in each cycle depends upon the design of the impulse valve and the magnitude of the delivery head. To allow for the effect of air in the system, an approximate average value for the acoustic velocity in the steel drive pipe of a typical ram would be in the region of 1000 to 1100 m/s. Some ram pump designs, such as the Sano, have impulse valves which operate under water and therefore do not ingest air.

Early history of the Ram Pump

John Whitehurst, a clock-maker and Fellow of the Royal Society, is credited with the invention of the first version of the hydraulic ram pump in 1772. Whitehurst had observed that when a domestic water tap is suddenly turned off, significant pressures were produced, sometimes high enough to burst the supply pipe. He devised a machine (see Figure 5.3) to harness this energy and used it to pump water from the basement to the top floor of a brewery at Oulton in Cheshire. The pump was not automatic and had to be activated by a child who opened and closed a tap. Nevertheless, this rather crude device must have been reasonably successful since it is said to have remained in operation until 1800.

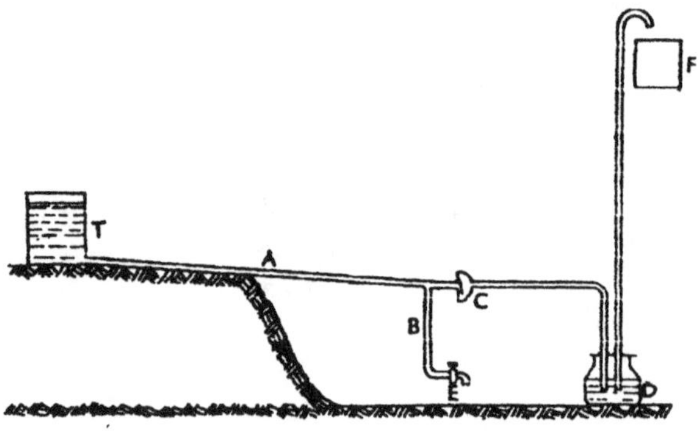

Figure 5.3 Whitehurst's machine

Joseph-Michel Montgolfier devised a significant improvement to Whitehurst's machine — the automatic valve. Joseph-Michel called his machine *le belier hydraulique*, the hydraulic ram. In the years which followed, many attempts were made to improve on Montgolfier's design but few, with the possible exception of the addition of a "snifter" valve and the rubber disc impulse valves used in the Vulcan and Blake rams, have stood the test of time. The ram pump today is basically in the same form as it was when Montgolfier invented it in 1776.

An historical review of experimental and analytical studies on ram pumps, which is well worth consulting, has been given by Rennie and Bunt (1981a) who describe work dating from as early as 1805.

Design

The following design procedure for ram pump systems with galvanised steel drive pipes is derived from Young (1998).

To start with, it is necessary to know the quantity q (in m^3/s) to be delivered, the height h (in m) of the delivery tank above the pump (in other words, the height to which the water is to be lifted) and the capacity of the source Q (in m^3/s).

For rural communities, the minimum q in litre/minute (L/min) = (number of people served)/20. To convert q to m^3/s, divide L/min by 60,000. The height h is obtained by adding together the measured, or physical lift, and the friction head loss in the delivery pipe (both in m). The source capacity Q (in m^3/s) may be determined using a bucket of known volume and a stop watch.

The next step is to calculate X, the characteristic head parameter. This is given by;

$$X = h/(H_s)^{1/2} \ (m^{1/2})$$

We know h and we must now calculate H_s, the static head of water needed to close the impulse valve. If the valve is of the conventional poppet-type, H_s (in m) is determined from:

$$H_s = M_v/\rho A_{disk}$$

where

ρ is the density of water (1000 kg/m³),

M_v is the mass of the moving parts of the valve (in kg) and

A_{disk} is the cross-sectional area of the valve disk (in m^2).

Example 1:

For a particular ram pump, the mass M_v of the moving parts of the impulse valve is 1.5 kg and, if the diameter of the valve disc is 62 mm, $A_{disk} = 3.02 \times 10^{-3}$ m^2. Thus $H_s = 1.5/(1000 \times 3.02 \times 10^{-3}) = 0.5$ m. If $h = 65$ m for example, $X = 65/(0.5)^{1/2} = 92.2$ $m^{1/2}$

We need to determine the following system parameters:

- D - the drive pipe diameter (m)
- H - the supply head (m)
- L - the length of drive pipe (m)
- n - the beat frequency (Hz)

but first we have to decide on a numerical value for three characteristic pump parameters, Z_1 (m/s), Z_2 (s) and Z_3 (s). These parameters are dimensional and may be represented collectively by Z_n where

$$Z_1 = nLh/H \text{ (m/s)},$$
$$Z_2 = HD^2/q \text{ (s)}$$
and
$$Z_3 = 2hD^2/Q \text{ (s)}.$$

Note that for any pump, $Z_n = Z_1 = Z_2 = Z_3$.

The conditions for a feasible ram pump design may be summarised as follows;

Condition 1:
For $X < 62$ m$^{1/2}$: $1.13\,X \leq Z_n \leq 6.2\,X$ (s or m/s)

Condition 2:
For $62 \leq X \leq 192$ m$^{1/2}$: $1.13\,X \leq Z_n \leq 385$ (s or m/s)

A lower bound for L has been recommended as 5m (Young 1995) and experience has shown that preferred beat frequencies lie between 0.5 Hz (30 per min) and 1.33 Hz (80 per min).

Example 2:

Using the value of X (= 92.2 m^{½}) calculated in Example 1, we find that Condition 2 applies for Z_n that is:

$$104.2 < Z_n < 385 \text{ (s or m/s)}$$

We can now choose any number between these limits to find: $D = (QZ_3/2h)^{½}$; $H = (q\,Z_2/D^2)$ *and* $L = (HZ_1/nh)$. *Select a value of* n *between 0.5 and 1.3 Hz.*

Example 3:

8 L/min is to be delivered to a height of 20m through a 32mm nominal diameter polyethylene pipe of length 5km. The friction head loss is estimated as 21m and the total delivery head to be provided by the pump is therefore 41m. The capacity of the source is measured as 200 L/min.

Solution:

For H_s = 0.5m, X = 58 m^{½}and, as X < 62 m^{½}, 65.5 < Z_n < 360 s or m/s.

Choosing $Z_n = Z_3$ = 150s for example, we obtain D = 0.078m. We will adopt the nearest lower commercial size of 0.076m for D. Then, with $Z_n = Z_2$ = 150s, H = 3.46m which is acceptable. Finally with $Z_n = Z_1$ = 150 m/s, L = 15.8m for a beat frequency of n = 0.8 Hz (48 per min).

The required beat frequency is set on installation by suitable choice of impulse valve stroke. If necessary, the valve mass may also be adjusted since this is not critical: a fairly wide range of choices for H_s is available without significant effect on the design.

Design Exercise:

Repeat the design of Example 3 for Z_n = 70, 100, 200, 250, 300 and 350 s or m/s. The answers are shown in Table 5.1.

Table 5.1 Answers to Design Exercise

Z_n (s, m/s)	D (mm)	H (m)	n (Hz)	L (m)	Q (L/min)
70	51	3.59	0.50	12.3	183
100	51	5.13	0.80	15.6	128
200	76	4.62	1.10	20.5	142
250	76	5.77	1.30	27.1	114
300	102	3.84	1.20	23.4	171
350	102	4.49	1.33	28.8	146

Discussion

Ram pumps are not difficult to design and install. Previous design methods have relied on arbitrary assumptions for the efficiency and irrational rules-of-thumb which often result in inoperable solutions. This has led to mistrust of this unique device and the impression that they are temperamental and "difficult". The research which has been carried out over the last five years at the Department of Civil Engineering of the Papua New Guinea University of Technology has resulted in a new understanding of the behaviour of ram pumps and the development of reliable design methods which work.

Papua New Guinea has the ideal terrain for the application of ram pumps which supply water from a fast flowing stream or waterfall to communities at a higher elevation. Typically, lifts are between 50 and 120m. The device is not expensive and is powered from the energy of the water itself (a renewable resource). There is no pollution of the environment and virtually no running costs apart from very occasional maintenance such as the replacement of valve rubbers. To illustrate this point, there is the case of a ram pump in England which ran for more than 100 years without any maintenance at all.

A glance at Table 1 illustrates a design strategy: the smaller the value chosen for the magnitude of the characteristic pump parameter Z_n (consistent with the topography of the site), the lower the overall cost of the pump system. Economic design also implies that reticulation systems are designed for least-cost.

The Civil Engineering Department has been involved in this area also and would be pleased to give advice on looped and branched water distribution network design as well as the design of the pump systems themselves. A recent example of the integrated design of a ram pump and associated branched reticulation network will be found at Gunemb settlement near Wau. The Department, in partnership with the Appropriate Technology and Community Development Institute (ATCDI), designed and installed a small system in 1996 to serve a population of about 75 people. Labour was supplied by the villagers and management of the project was the responsibility of two British VSOs working with ATCDI.

There were two unexpected social benefits for the settlers. For some years they had been fighting an eviction order issued by Wau Town Council for allegedly polluting the stream with detergents; this order has now been rescinded. The second benefit came from the local provincial government member who turned up at the opening ceremony with a donation of 200 Kina. He was inspired, no doubt, by the recognition and publicity given to the community by the involvement of the University and the fact that 1997 was an election year. The pump is still performing well and has not needed any attention since installation.

The Civil Engineering Department produces three sizes of ram pumps (25mm, 50mm and 80mm). These are manufactured in the University's Central Engineering Workshop from locally available materials and are therefore relatively inexpensive compared with imported items. The Department would be happy to respond to inquiries.

It is the author's hope that this short introduction to ram pumps will lead to greater use of the device in Papua New Guinea thereby contributing to national development and a better quality of life in rural communities.

Acknowledgements

The author would like to thank all the organisations and individuals who have assisted him with this research over the years. In particular, the Research Committee of the Papua New Guinea University of Technology; the technical staff of the Civil Engineering Department and the Central Engineering Workshop and, last but not least, the numerous students who carried out experiments on ram pumps as part of their Final Year Project requirements.

REFERENCES

Anderson, E. W. (1922) Hydraulic rams. *Proceedings of the Institution of Mechanical Engineers*, v102, 337-353

Azoury, P. H., Baasri, M. and Najm, H. (1988) On the optimum utilisation of water hammer. *Proceedings of the Institution of Mechanical Engineers, Journal of Power Engineering*, 202, 249-256.

Calvert, N. G. (1957) The hydraulic ram. *The Engineer*, 203(5282), 597-600.

Calvert, N. G. (1958) Drive pipe of a hydraulic ram. *The Engineer*, 206(5370), 1001.

Chiang, Y. C. and Seireg, A. A. (1985) Hydraulic rams - computer simulation and optimum design. *Computers in Mechanical Engineering*, 3(4), 65-68.

Fyfe, W.W. (1922) *Automatic Pumping*. Charles Griffen.

Harralson, J. H. and Cozens, P. (1982) Numerical simulation of the hydraulic ram: a new look at an old device. *Proceedings of the 28th Heat Transfer and Fluid Mechanics Institute*, 91-104.

Iversen, H. W. (1975) An analysis of the hydraulic ram. *Transactions of the American Society of Mechanical Engineers, Journal of Fluids Engineering*, Ser. 1, 97(2), 191-196.

Krol, J. (1951) The automatic hydraulic ram. *Proceedings of the Institution of Mechanical Engineers*, v165, 53-65.

Lansford, W. M. and Dugan, W. G. (1941) An analytical and experimental study of the hydraulic ram. *University of Illinois Bulletin*, 38(22).

Li, W-H. (1983) *Fluid mechanics in water resources engineering,* Allyn and Bacon, 237.

Mead, D.W. (1933) *Hydraulic Machinery*, McGraw-Hill, 358-383.

O'Brien, M. P. and Gosline, J. E. (1933) The hydraulic ram. *University of California Publications in Engineering*, 3(1).

Philip, A. and Harralson, J. H. (1991) Experimental validation of the hydraulic ram simulation model. *Proceedings of the 32nd Heat Transfer and Fluid Mechanics Institute,* 109-140.

Rankine, W. J. M. (1872) On the mathematical theory of the hydraulic ram. *The Engineer,* v33.

Rennie, L. C. and Bunt, E. A. (1981a) The automatic hydraulic ram. Part 1: *South African Mechanical Engineer*, v31, 258-273.

Rennie, L. C. and Bunt, E. A. (1981b) The automatic hydraulic ram, Part 2: *South African Mechanical Engineer*, v31, 286-311.

Rennie, L. C. and Bunt, E. A. (1990) The automatic hydraulic ram - experimental results *Proceedings of the Institution of Mechanical Engineers, Part A: Journal of Power and Energy,* v204, 23-31.

Schiller, E. J. and Kahangire, P. (1984) Analysis and computerised model of the automatic hydraulic ram pump. *Canadian Journal of Civil Engineering,* v11, 743-750.

Slack, D. C., Eshenaur, W. C. and Berkas, T. H. (1984) Predicting the performance of a water-pumping hydraulic ram. *International Journal for Development Technology*, v2, 261-271.

Tacke, J.H.P.M. (1988) *Hydraulic rams - a comparative investigation. Rep. No. 88-1*, Delft University of Technology.

Verspuy, C. and Tijsseling, A. S. (1993) Hydraulic ram analysis. *Journal of Hydraulic Research,* v31(2), 267-278.

Young, B. W. (1995) The design of hydraulic ram pump systems. *Proceedings of the Institution of Mechanical Engineers, Part A: Journal of Power and Energy,* v209 (A4), 313-322.

Young, B. W. (1996a) System designs for the Platypus ram pump, *Agricultural Engineering Australia,* v25, n1, pp13-18.

Young, B. W. (1996b) Sizing ram pumps. *Proceedings of the Institution of Mechanical Engineers, Part A: Journal of Power and Energy,* v210 (A4), 245-248.

Young, B. W. (1996c) Simplified analysis and design of the hydraulic ram pump, *Proceedings of the Institution of Mechanical Engineers, Part A: Journal of Power and Energy,* v210 (A5), 295-303.

Young, B. W. (1997) Design of homologous ram pumps, *Transactions of the American Society of Mechanical Engineers, Journal of Fluids Engineering,* v119, n2, 360-365.

Young, B. W. (1998) Generic design of ram pumps. *Proceedings of the Institution of Mechanical Engineers, Part A: Journal of Power and Energy,* v212 (A2), 117-124.

6

Positioning by Satellite – The Myths and Applications in the Environment

R.A. Curley

ABSTRACT

It is more than forty years since the launch of the first artificial satellites by both the USSR and the USA. The first generation navigation system, TRANSIT, soon followed these launches but was almost exclusively used by surveyors, geodesists and the military due to the cost and size of the equipment and the esoteric nature of the observations and processing required. The current system, Navstar GPS, brings positioning by satellite (GPS) within reach of just about anybody who needs to know his position on or above the surface of the earth using equipment the size and price of a scientific calculator and the user needs no specialist knowledge to operate the equipment. The applications of GPS are limited only by the imagination and innovation of the user and GPS receivers seem destined to become as common and essential as mobile telephones, fax machines and the computer for many professionals. This paper looks at the development of GPS and its application in the environment with particular reference to forest covered areas of the world such as Papua New Guinea.

Introduction

Much has been written about GPS since its conception in the 1970s and there are a number of excellent texts giving technical details of the system and the characteristics of the GPS signal structure (Spilker 1986). Therefore it is considered unnecessary to repeat the details in this paper. However, it is important to emphasise the intended use and the capabilities of the system and, therefore, a brief history of positioning by satellite and a description of Transit Doppler and GPS is considered appropriate.

The first artificial satellite, Sputnik 1, was put into space by the former USSR on 4 October 1957 and tracking stations around the world monitored the progress of the satellite as it orbited the earth. With a knowledge of the co-ordinates of the tracking stations and the distance or range measurements to the satellite, the parameters of the satellite's orbit could be determined and hence it was possible to calculate the position of the satellite in its orbit at any given moment in time. It followed that the reverse process should also be possible, that is knowing the orbital parameters and hence the position of the satellite in its orbit and the range to the satellite, it should be possible to determine the position of the receiver taking the measurements.

This was the basis of Transit Doppler, the doppler shift of the satellite signal observed at the receiver providing the means of determining the range to the satellite and hence the position of the receiver. The portable satellite receivers used to record the satellite signal were expensive, $US55,000 in the 1960's. There were few commercial models of receivers available, perhaps the most notable being the Magnavox and JMR models. Apart from the cost of the receivers other factors prevented the widespread use of the system, it took a long time to get an accurate fix, in some case as much as three days to get a fix with an accuracy within the 1-2 metre range. Also the observables and the post-processing techniques were incomprehensible to all but the specialists. Hence the use of Transit Doppler was largely restricted to government survey agencies and large international corporations engaged on expensive exploration surveys in remote parts of the world where conventional survey techniques proved slow and inadequate.

The GPS system

The system is generally considered to comprise three segments:

- The Space Segment
- The Control Segment
- The User Segment

It is important to realise that users do not have any control of either the Space Segment or the Control Segment both of which are operated by the U.S. Department of Defence.

The Space Segment consists of 24 operational satellites in six orbital planes inclined at approximately 60 degrees to the equator. The satellites orbit at a height of 20,000 kms and each satellite, with a velocity of approximately 4 kms per second, completes an orbit every 12 hours. The inclination of the satellite orbits, the altitude of the satellites and the position of individual satellites in each of the orbital planes were carefully chosen in order to provide the most cost-effective way of providing continuous world-wide coverage.

The Control Segment is responsible for putting the satellites into space and positioning them in their orbits. The Control Segment tracks and monitors the behaviour of the satellites and its component parts and uplinks information to the satellites for broadcast by the satellites to the users.

The User Segment includes any person, vessel, aircraft, vehicle, etc., with equipment capable of receiving and decoding the satellite signal for the purpose of providing positional, navigational or timing information. Whilst the Control Segment and the Space Segment remain the same, except for replacement of satellites, the User Segment is increasing and diversifying at an amazing rate.

Brief History and Present Status of the GPS System

The following is an edited extract from the Federal Radionavigation Plan 1986 (FRP 86) under the heading of Revised Comprehensive GPS User Policy (effective 22 May 1985)

The GPS is a continuous world-wide satellite-based radio navigation system currently in full-scale development by the DOD, the system will provide properly equipped users the capability to obtain navigation and geodetic positions in three dimensions, velocity in three dimensions plus highly accurate time. The system will simultaneously transmit navigation information that permits positioning accuracy at two levels. The higher level of accuracy can be obtained from the Precision Positioning Service (PPS). Reduced precision is available through the Standard Positioning Service (SPS).

This policy provides that the DOD intends that the SPS signal will be broadcast in the clear and will be available for use by any properly equipped user. There will be no annual or other direct fee associated with the use of this signal. The SPS will be made available to civil, commercial and other users on an international basis at the highest level of accuracy consistent with U.S. National Security interests.

The PPS signal will be encrypted and will be made available initially to US and selected allied military users. Limited civil use of the PPS may be authorised if it can be demonstrated that such use is in the national interest, adequate security protection can be provided, and comparable accuracy cannot be obtained from another source. Indirect costs, such as nonrecurring cost recoupment and purchase of the decryption and anti-spoof devices may be applicable to use the PPS.

This policy statement makes the following points quite clear:

- GPS is a military navigation system not a civil surveying system.
- Civil and commercial users will have free access only to the SPS
- Use of the PPS is restricted to the military and may only be extended to other authorised users at the discretion of the DOD

US policy has not changed since this statement was issued in 1985 despite the civil and commercial users increasing reliance on GPS as a primary navigational system and the lobbying for the PPS to be declassified.

GPS was implemented in three phases beginning with Phase I and concept validation from 1973 -79. During this phase test satellites were launched to provide a minimum configuration to permit testing of prototype user equipment and to prove the viability of the system in general. The first satellite was launched on the 23 February 1978 and the minimum four satellite configuration, or constellation, was completed by January 1979. The constellation provided up to four hours per day of suitable satellite coverage over a test site in the USA. Four contractors were invited to develop military user equipment for validation and testing.

Phase II, the fullscale development of the system took place during the period 1979-86. This phase verified the operational effectiveness of the GPS concept and helped identify problems with the space and user segments of the system. Many changes were made to both hardware and software designs and to the specifications of the satellites and the receivers. During this phase satellites were replenished on an as need basis so that six satellites were available throughout the testing period. The satellites launched during this phase were referred to as Block I satellites or space vehicles (SV's). The design life of the satellites was 7.5 years so there are no Block I satellites currently operating.

It was during this phase that many civil and commercial users began to get involved and started using the system for surveying, geodetic and exploration surveys and many universities around the world began research into the use of GPS and methods on the avoidance of the restrictions on the accuracy imposed on the SPS. Many papers were published during this period debating the pros and cons of various receiver architectures, use of the various observables and other contentious issues. Some of these issues are still being debated but the large majority have been settled with the passage of time.

Phase II was followed in turn by Phase III, the production, development and emplacement period from 1986 to 1994. This phase began when in 1983 Rockwell were awarded the contract to build twenty-eight Block II satellites and production of military user equipment began in 1986. The satellites were launched on both the Space Shuttle and expendable launch vehicles (ELV's) and several problems besieged this phase, perhaps the most disastrous was the explosion of the space shuttle in January 1986. The first Block II satellite was launched in 1989 and the system officially achieved initial operational capability (IOC) in December 1993 with a 24 satellite constellation consisting of a mix of both Block I and Block II satellites. It is only at this time that the service available to the civil user was guaranteed at the SPS level.

The current GPS constellation consists of 25 Block II/IIA satellites operating with full operational capability (FOC). The initial Block II satellites were modified and these are referred to as Block IIA satellites, the first such satellite was launched in 1990. Since the first of the Block II satellites was launched in 1989 and they had a design life of 6.5 years, replacement satellites have been launched. These satellites are referred to as Block IIR satellites and the first of these, SV43, was launched successfully on 22 July 1997 after a failed attempt in January of the same year.

System Description

The satellites are in near circular orbits, at an altitude of 20200 km, with an approximate 12-hour period. The geometry of the satellites thus repeats itself each sidereal day and appears, to a static user, to be in the same position in its orbit four minutes earlier each day.

Each satellite transmits two L-band frequencies 1575.42 Megahertz (L1) and 1227.6 Megahertz (L2). The L1 frequency transmits both the PPS (P-Code) and the SPS (Course Acquisition or C/A Code). The L2 frequency transmits the P-Code only. The P-Code is generated at the GPS clock frequency of 10.23 Megahertz. One element of the code sequence corresponds to a time interval of about 100 nanoseconds which is equivalent to 30 metres in distance, or range, which is the more usual term used in navigation. The P-Code does not repeat itself for 267 days. The 267-day sequence is subdivided in such a way that each satellite is given a unique one week portion of the code that does not overlap with that of any other satellite. The C/A-Code is less complex and has a frequency of 1.023 Megahertz corresponding to a range measurement of 300 metres. Since it repeats itself every millisecond and the speed of light is approximately 300,000 km/sec this implies an integer ambiguity of multiples of 300km. Both codes have characteristics of random noise but are in fact binary codes generated by mathematical algorithms and are thus called "pseudorandom" noise codes (PRN). These codes identify each satellite in the constellation and the satellite is usually referred to by its PRN number.

Superimposed on these signals are navigation and system data including satellite ephemerides, atmospheric propagation correction data and satellite clock bias information to correct each satellite to GPS system time. The satellite ephemeris is a set of parameters that allows for the computation of the predicted position of the satellite in its orbit with respect to time. Since these parameters are broadcast as part of the satellite message the ephemeris is referred to as the broadcast ephemeris. The observed position of the satellite in its orbit can be obtained from a precise ephemeris that is produced after the orbit is completed.

The Basic Navigation Principle

In its design mode, that of near instantaneous positioning and navigation, the GPS concept is based on accurate and continuous knowledge of the position of each of the satellites in the system with respect to time and the measurement of the time of arrival, at the receiver, of the signals transmitted by each satellite. The difference **(Dt)**,between the transmit time (t_T) of the signal from Satellite (i) and the time of receipt of the signal, at the receiver position **(A)**, (t_R) when scaled by the speed of light (C) gives the range (R) from the receiver to the satellite (Fig. 6-1).

$$(t_R)_A - (t_T)^i \times C = \Delta t_A^i \times C = R_A^i$$

The range from the receiver **(A)** to the satellite (i) can also be determined from the co-ordinates of both the receiver and the satellite which are expressed in 3-D geocentric cartesian co-ordinate system (Fig. 6.2).

$$R_A^i = \sqrt{(X_A - X^i)^2 + (Y_A - Y^i)^2 + (Z_A - Z^i)^2}$$

Equating (1) and (2) gives

$$(t_R)_A - (t_T)^i \times C = \Delta t_A^i \times C = R_A^i = \sqrt{(X_A - X^i)^2 + (Y_A - Y^i)^2 + (Z_A - Z^i)^2}$$

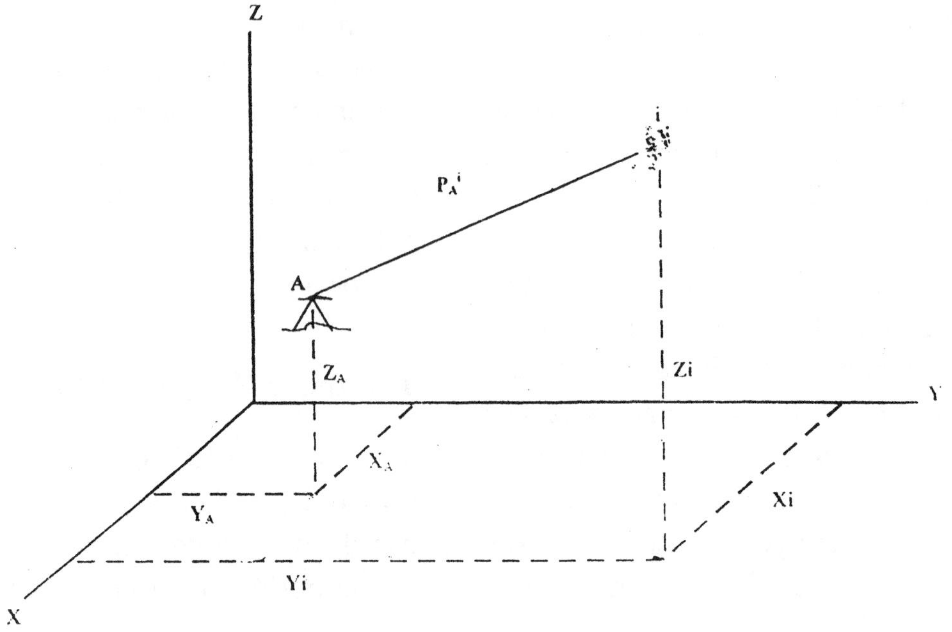

Figure 6.1 Satellite to receiver range

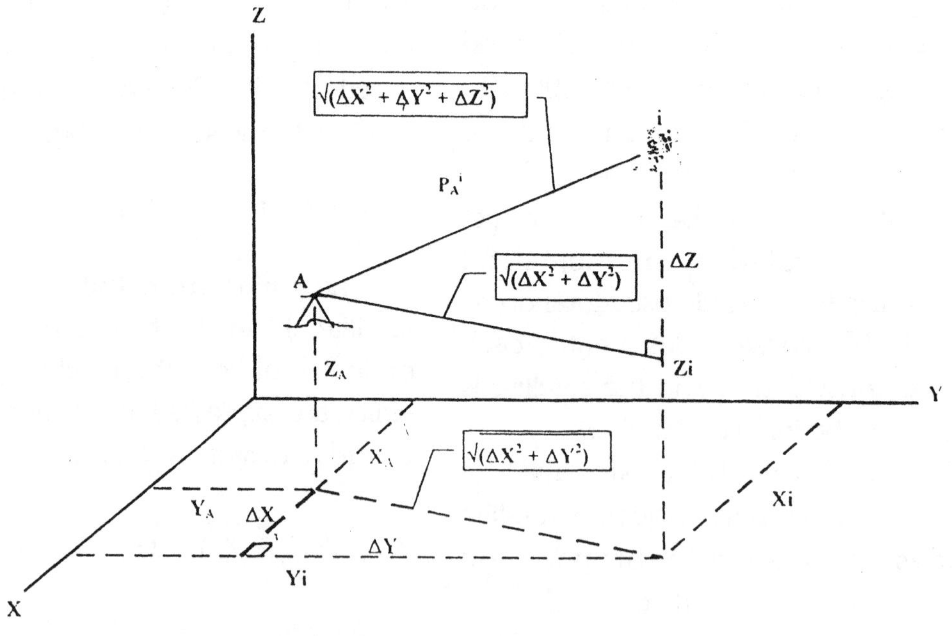

Figure 6.2 Range from 3-D cartesian coordinates

The measurement of three ranges to three satellites (i,j,k) would produce three simultaneous equations with three unknowns, the receiver X,Y and Z, which can be solved in the normal way.

Since the satellite clock is used to obtain the transmit time of the signal and the receiver clock is used to obtain the arrival time of the signal and these may not be synchronised a fourth range to an additional satellite is observed to determine this unknown local receiver clock offset. (E).This results in four equations of the form:

$$\Delta t_A^i + \mathbf{E} \times C = R_A^i = \sqrt{(\mathbf{X}_A - X^i)^2 + (\mathbf{Y}_A - Y^i)^2 + (\mathbf{Z}_A - Z^i)^2}$$

$$\Delta t_A^j + \mathbf{E} \times C = R_A^j = \sqrt{(\mathbf{X}_A - X^j)^2 + (\mathbf{Y}_A - Y^j)^2 + (\mathbf{Z}_A - Z^j)^2}$$

$$\Delta t_A^k + \mathbf{E} \times C = R_A^k = \sqrt{(\mathbf{X}_A - X^k)^2 + (\mathbf{Y}_A - Y^k)^2 + (\mathbf{Z}_A - Z^k)^2}$$

$$\Delta t_A^l + \mathbf{E} \times C = R_A^l = \sqrt{(\mathbf{X}_A - X^l)^2 + (\mathbf{Y}_A - Y^l)^2 + (\mathbf{Z}_A - Z^l)^2}$$

In essence the receiver computes a position based on the intersection of the four measured ranges to the four satellites whose positions are given in the satellite ephemeris data which is broadcast as part of the satellite message. The position is calculated in the same reference system as the satellites and tracking stations which is the World Geodetic System 1984 (WGS84). If more than four satellites are used then this provides a check on the solution. These positions can be transformed to fit local mapping if a suitable set of transformation parameters exist (Fig. 6-3).

The accuracy of the determined position is dependent on a number of factors (Fig. 6.4) which includes: residual receiver and satellite clock errors, propagation errors along the signal path through the ionosphere and troposphere, satellite position or orbit errors, measurement errors, multipath errors and errors due to unfavourable satellite geometry. But by far the

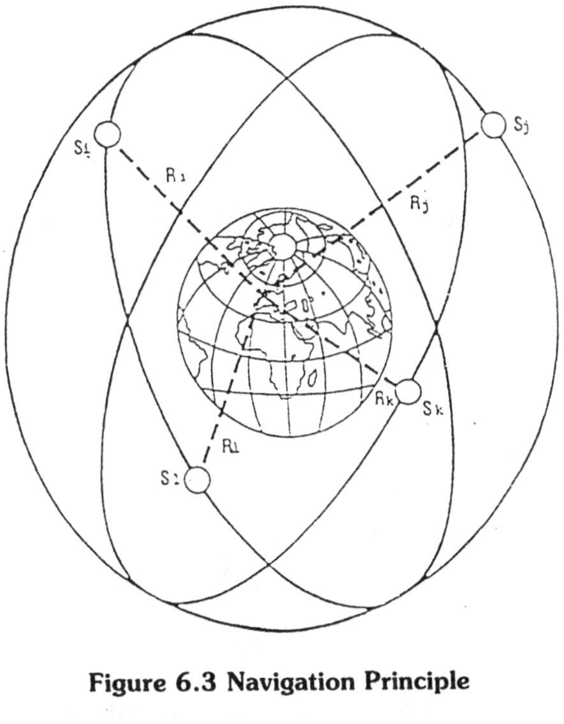

Figure 6.3 Navigation Principle

largest source of error is the deliberate imposition of errors by the Control Segment in the form of Selective Availability.(S/A).

This method of positioning is called point positioning or pseudoranging, this term is used because it is not the true range that is used in the positioning calculation but the observed range which is biased by the effects of the factors listed in the previous paragraph. The quoted accuracies (2 sigma) in this single receiver mode are given in the FRP (1986) as:

Using the SPS or C/A Code	100.00m	Horizontally
	156.0m	Vertically
Using the PPS or P-Code	17.8m	Horizontally
	27.7m	Vertically

It is generally accepted that SPS will give positions accurate to within 100m 95 per cent of the time and since unauthorised civilian users only have access to the SPS this is the best that can be achieved with a single receiver in point positioning mode. It should be better than 300m 99.9 per cent of the time.

Selective Availability

The deliberate degradation of the positioning capabilities of the C/A code is referred to as Selective Availability (SA) and it is the removal of this restriction on the system that is the main focus of the lobbying by the civilian satellites

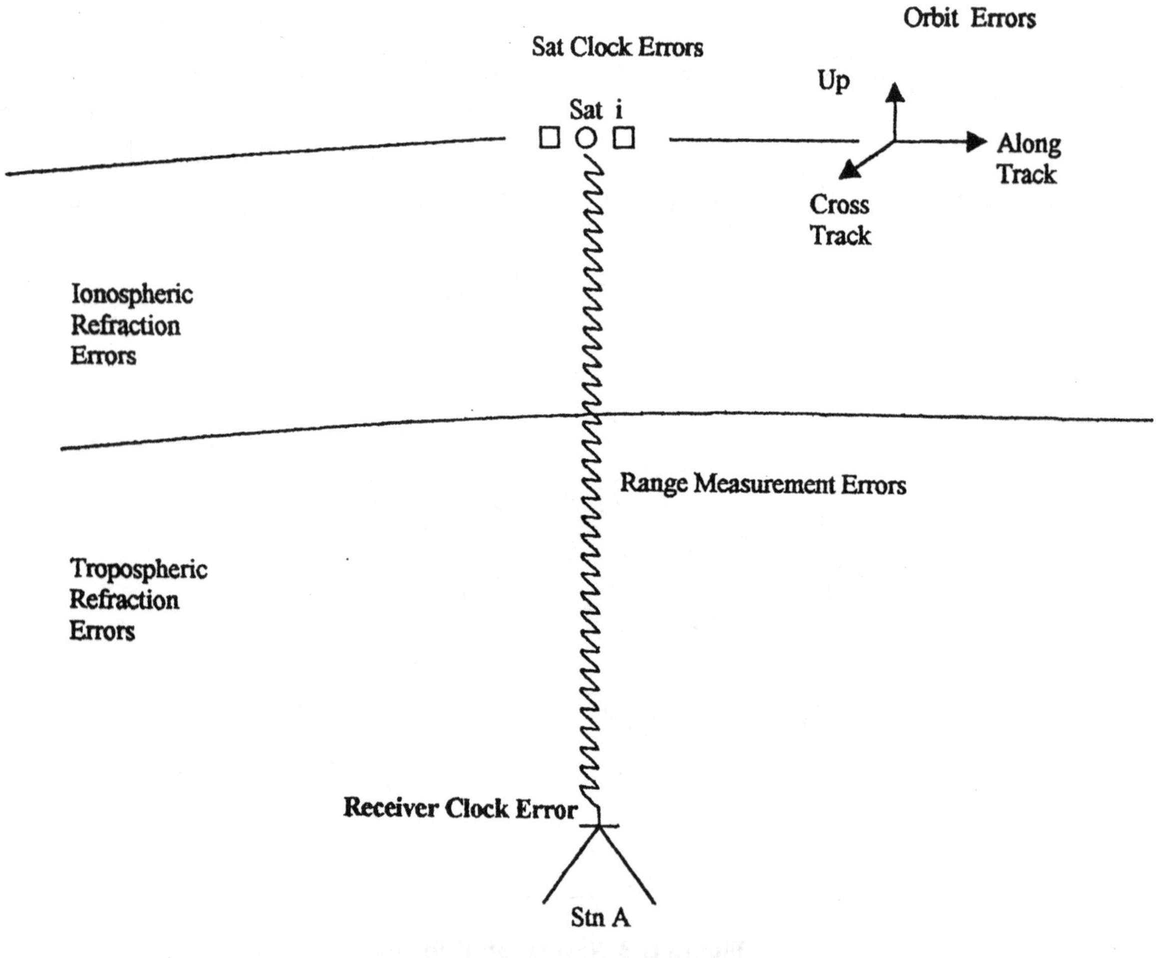

Figure 6.4 Errors in the Pseuodorange

user groups. The US government has indicated that it will remove this restriction some time in the future but no definite date has been given as to when this will happen.

Improving on the SPS Accuracy

To get round this problem of S/A non-authorised users of the P-Code employ a technique of differential positioning, usually referred to as DGPS. This technique requires a minimum of two receivers, one, the master or reference, is set up over a known base station and the other, the remote or rover, is taken to each of the points to be fixed. Both receive the C/A Code from at least four common satellites. If results are not required in real time then the data can be recorded in both receivers and later downloaded into a computer for postprocessing. The observed data at the master and the known position of the master are used to determine corrections to be applied to the observed data collected by the rover on each of the unknown sites.

If results are required in real time then the differences in the known and observed positions at the master station are used to calculate corrections to the pseudoranges to each of the satellites and these corrections are then transmitted to the rover and used to correct the observed position. The displayed position then has the major effects of S/A removed and as a consequence the positional accuracy is improved to the level of the PPS, or better, depending on the sophistication of the differential technique used. This technique can be used to obtain decimetre positional accuracy in certain circumstances but potential users of this technique should be wary of advertising literature and make sure they understand the receiver configuration and differential method being used. Important factors are, the number of reference stations being used to calculate the corrections, the method and rate of transmission of the corrections, the integrity and quality control of the transmitted corrections.

It is possible for a user to set up his own reference station and, with suitable firmware and a radio transmitter, calculate and transmit corrections to any roving field receivers. The area that can be worked over will depend on the capabilities of the radio hardware. This method is used extensively in seismic exploration surveys for setting out receiver and source positions. Commercial companies have set up differential networks covering most of the globe, in the Pacific and Australasian Regions both Omnistar and Landstar are operation and can be used in Papua New Guinea. (Roberts 1997).

Receiver Types

GPS is already a multi-billion dollar industry. Recent estimates predict that the industry will be employing some 100,000 people and be generating 6 billion US dollars by the end of the decade. The contract for supplying the follow-on satellites due to be launched in 2001 is reputedly worth 3.1 billion US$.

In January 1990 *GPS World* was launched as a magazine dedicated solely to GPS topics. Current circulation is over 40,000 copies world-wide and it has grown into an authorative journal on everything to do with GPS. The growth of the GPS industry is probably best illustrated by quoting some statistics from recent issues of the magazine. The 1997 receiver survey reveals that 61 manufacturers of GPS receiver equipment produce 394 different models of receiver. The receiver types fall into broad user categories including recreation, aviation, survey/GIS, marine, timing defence, vehicle tracking, real time DGPS, navigation and others.

It is beyond the scope of this paper to make recommendations on receiver/software configurations. Suffice to say that whatever the requirements, there is sure to be a system available to meet the needs of the user and if there is not then manufacturers are keen to develop and adapt current hardware and software to suit their clients' needs. Many suppliers actually sell solutions to positional problems rather than just receivers.

The Myths

The myths surrounding GPS arise mainly from misunderstanding the basic concepts of the system. Although it is not necessary to understand geodesy to switch on a receiver and get a position, for example, a knowledge of basic geodesy is helpful when you want that position in a particular co-ordinate system or when you want heights above mean sea level. GPS requires line of sight between the receiver and the satellites and provided this can be achieved it can be used in any environment for any task that requires spatially referenced data. If the receiver is mounted on a platform that is moving rapidly special antennae may be required (Schulper & Clark 1991), but the system is designed to cope with all military scenarios so it should work in most civilian applications. Algorithms have been developed to cope with momentary loss of signal, when passing under a bridge for example. Likewise it is possible to use GPS in tree covered areas but this might require special antennae also (see below).

With S/A switched on it is only possible to obtain a position accurate to 100m when using a single standalone receiver, regardless of the type of receiver. The SPS guarantees this level of accuracy 95 per cent of the time. The other 5 per cent of the time it is possible for the position to be several hundreds of metres out. To ensure positions accurate to the 10-metre level it is necessary to use some form of differential technique.

The heights obtained from GPS relate to heights above the surface of the spheroid used to define the position of the observer. Generally what is required is heights above mean sea level. Although it is possible to convert between the two systems of heighting, if the separation between the two surfaces is known, in reality this proves difficult in most parts of the world, in particular places like Papua New Guinea. The main reason for this is because the reference surface for heights above mean sea level is the geoid which is an irregular surface. Therefore, the value for the separation changes from place to place sometimes by several metres. In Papua New Guinea there is a variation in this surface of about 30 metres. Attempts have been made to provide models of this surface on a global, national and local basis but clear understanding of these concepts goes beyond the casual users knowledge. Suffice to say that it is possible to obtain accurate mean sea level heights from the system if an accurate geoidal model exists. Models for Papua New Guinea do exist and the most recent, available from the Papua New Guinea National Mapping Bureau (NMB), should provide relative mean sea level heights accurate to 2 or3 metres over small areas. (Curley & Jackson 1997).

GPS Applications

Essentially GPS is just a positioning and navigation system. What makes it different from other systems is that, it is easy to use, it is instantaneous, it is available world-wide, it is continuous 24 hours a day in any weather and is absolutely free once the receiver has been purchased. Obvious uses are surveying and mapping and straightforward navigation. However it should be pointed out that the application of GPS for surveying and geodesy, where accuracies of a few millimetres are achievable, needs special equipment and procedures which is beyond the scope of this paper (Curley 1998a). Satellite positioning was first used in Papua New Guinea by the Royal Australian Surveying Corps in 1974. Transit Doppler was used extensively between 1974 - 79 to fix the positions of 202 geodetic survey stations, 151 of which were on the mainland and 51 on surrounding islands (Laing 95). Transit Doppler was used to determine plate tectonic movement in 1981 (Stoltz 1989). The first published Papua New Guinea GPS survey was carried out in 1990,to monitor crustal motion and was repeated again in 1991 and 1992 (McClusky 1994).

More of these type of surveys were carried out by visiting earth scientists from the University of Santa Cruz (1993, 1995, 1997 and 1998) and the Australian National University in (1996, 1998). The NMB, the Papua New Guinea University of Technology, the University of Papua New Guinea and the Geological Survey Office were all involved in these joint campaigns. The NMB obtained their first GPS receivers in 1993 but their use has been restricted to providing land survey control. They acquired a code only receiver to carry out kinematic surveys of roads around the capital district while they have continued to use their geodetic receivers to carry out surveys of all the airstrips in Papua New Guinea (Rosa 98). The Rabaul Volcanic Observatory is establishing a fixed array of survey grade GPS receivers for the continuous realtime monitoring of volcanic activity.

The private survey sector in PNG also became aware of the potential of GPS and some purchased receivers for control work and later made use of both the handheld and G.I.S. models (Pratt 97). Most fishing boats in Papua New Guinea waters now use GPS routinely as their primary navigation source but to the author's knowledge none of the Yacht Clubs in Papua New Guinea provides a DGPS service for their members. In 1997 DGPS was used by Terra Surveys, a Canadian company, to provide a base map of the Huon Gulf for future geo-science studies. Both the University of Technology and the NMB operate continuous base stations which can be used for DGPS, but at the time of writing neither provide a real time differential service.

Broader Applications

The applications of GPS appear limitless:

- It has been used to monitor the grazing habits of sheep in Wales, following the Chernobyl accident, to determine where the sheep were grazing and to correlate this with the individual sheep's radiation level to try and locate likely hotspots of radiation (GPS World Showcase 1995)
- It has been used for mapping malaria in Kenya (GPS World Showcase 1995)
- It has been used for assessing damage from natural disasters (Bottorff 1994)
- It is used in the transport industry for tracking road vehicles (Hossfeld 1991) and on the railways for keeping track of trains (Gibbon 1990)
- It is used with inertial navigation systems in aviation (Diesel & Guerrazzi 1991)
- In hydrographic surveys with echo sounders to map siltation in rivers (Chisolm & Lenthall 1991)
- With Remote Sensing for shallow water mapping (Sipuman 1998)
- In agricultural land use studies (Perry 1992)
- In oceanography as a remote sensing tide gauge (GPS World Showcase 1996)
- It has been used as an asset management tool in forestry (Petersen 1990)
- It is used on civil engineering for plant to control earthworks.
- It has been used to measure volumes of stock piles in mining and quarrying.
- It has been used for precision farming (Bauer & Schefcik 1994).
- For monitoring coastal erosion (Byrnes et al 1994).

Future Applications

The future uses of GPS are limited only by the imagination and innovation or current and potential users. Digital Mapping and GPS are complementary. So digital maps on CD-ROM and GPS receivers can provide the basis for an in-car navigation system. Such systems are currently available for top of the range cars. It is sure to become standard on all models in the future in the same way that in-car entertainment is now. Another potentially large market is for transport fleet management. Vehicles will continuously relay their positions and their velocities to a control station. There are also security applications. A stolen vehicle can relay its location to the authorities which will assist the recovery of stolen vehicles and probably result in lower insurance premiums for those vehicles fitted with such devices. All Civil Aviation must switch to GPS and World Geodetic System 1984 (WGS84) in the year 2000 thus providing one global seamless map. All navigation aids, in and around airports, and runway details will be expressed in this common co-ordinate reference frame. Instrument landings will use DGPS.

In most hydrographic applications, data from other sources, for example, gyros for headings, are used in conjunction with GPS to provide primary navigation in mid-ocean as well as ports and harbours. In the recreation industry weekend yachtsmen, divers, fishermen, mountaineers, golfers, orienteers, and ramblers are just some of the sportsmen using GPS technology in leisure pursuits. It should be remembered that GPS not only provides position in 3 dimensions but also velocity in

3 dimensions, so it is ideal for keeping track of vehicles in motor sports.

The Russians have a similar system called GLONASS which is declassified and does not suffer from the problems associated with selective availability. Manufacturers are now making receivers that can track and make use of both GLONASS and GPS signals. This not only improves the accuracy of the positioning but provides more satellites to choose from in positions where a clear view of the sky is restricted. In forest areas, particularly equatorial rain forest areas, the GPS signal does not penetrate the tree canopy very well. However, special antennae have been constructed to try to improve reception and there have been many reported cases where GPS has been used successfully in such scenarios (Bergstrom 1990,Petersen 1990, Flaherty & Rowe 1997).

Conclusion

The specialist users such as surveyors and geodesists are well aware of the capabilities and limitations of GPS. Earth scientists adopted its use very early on as it provides the only means of obtaining accurate and repeatable measurements over large distances for plate tectonic and crustal motion studies. However, the word is spreading, through general papers such as this, and a much wider group of potential users is emerging. These are people who do not need millimetric or even sub-metre accuracies but require a position sufficient to spatially reference some attributes relating to the environment such as population details, land use, etc. The author has noted that, although GPS has only been available in the Department of Surveying and Land Studies at the University of Technology since 1995, interest and usage have already spread to the Forestry Department, Agriculture Department, Civil Engineering Department and the Environmental Research and Management Centre. The above references show the wide variety of uses of GPS in other parts of the world and many of these could be applied in Papua New Guinea; therefore the following comments are made:

- GPS is a freely available continuous worldwide navigation, positioning and timing system which requires little or no skill to use.
- There are many models of receivers available for various applications; the basic handheld navigation system would satisfy most non-surveying or geodetic uses.
- DGPS using the code only observable provides positioning accuracies of better than 10 metres and is capable of achieving sub-metre accuracy in certain scenarios.
- The applications of GPS appear limitless but the growth areas include vehicle tracking, leisure pursuits, the built and natural environment and the earth sciences.
- GPS technology can be combined with other technologies to provide integrated solutions to many positioning problems.

REFERENCES

Bauer, W.D. and Schefcik,M. (1994) "Using Differential GPS to improve Crop Yields" *GPS World.* Feb 1994: 38-41.

Bergstrom, G.C. (1990) "GPS in Forest Management" *GPS World,* Sep/Oct 1990: 46-49.

Bottorff, H. and MacSwan, K.J. (1994) "GPS helps assess Mississippi River Flood Damage" *GPS World* Mar 1994: 22-26.

Byrnes, M.R., McBride, R.A.,Underwood,S.G.,and Corbley,K.P. (1994) "Losing Ground: Mapping Louisiana's Disappearing Coastline" *GPS World* Oct 1994: 46-50.

Curley, R.A. (1998a) "GPS Field Procedures" *33rd Survey Congress, Association of Surveyors of Papua New Guinea,* Madang, July 1998.

Curley, R.A. (1998b) "Proposed amendments to the Survey Directions on Satellite Positioning Surveys in Papua New Guinea" *33rd Survey Congress, Association of Surveyors of Papua New Guinea,* Madang, July 1998.

Curley, R.A. and Jackson, R.(1998) "GPS Heighting – The Ups and Downs" *32rd Survey Congress, Association of Surveyors of Papua New Guinea,* Port Moresby, July 1997.

Chisolm, G. and Lenthall,P.(1991) "Combining GPS and Echo Sounders to Monitor River Siltation" *GPS World*, May 1991: 46-48.

Diesel, J. and Guerrazzi,E.(1991) "Synergistic integration of GPS and INS for Civil Aviation" *GPS World,* May 1991: 41-45.

Duffy, M.A.(1991) "GPS and the Cadastral Survey" *GPS World,* May 1991:38-40.

1986 Federal Radionavigation Plan. US Dept of Defence,Washington, D.C. Document DOD-4650 4 DOT-TSC-RSPA-87-3.

Flaherty, M. and Rowe,G. (1997) "The Use of New Technology for Surveying and Mapping in the Forest Sector" *The New Zealand Surveyor.* No.287 Apr 1997: 14-19.

Gibbons, G.(1990) "On Track with GPS" *GPS World* May/Jun 1990: 28-37.

GPS World Showcase. (1995) Vol 6 No 8 Aug 1995. Advanstar Communications: Oregon.

Hossfeld,B. (1991) "GPS for Vehicle Tracking" *GPS World* Oct 1991:44-46.

Kevie van der, J. and Gould,R. (1998). *33rd Survey Congress, Association of Surveyors of Papua New Guinea,* Madang July 1998.

Perry, E.M. (1992) "Using GPS in Agricultural Remote Sensing" *GPS World* Jul/Aug 1992: 30-36.

Petersen, C. (1990) "Into the Woods with GPS" *GPS World* Nov/Dec 1990:31-36.

Pratt, F.C. (1997) "Reflections of a Reactionary" *32nd Survey Congress, Association of Surveyors of Papua New Guinea.* Port Moresby, June 1997.

Roberts, W. (1997). "Commercial DGPS for Land Surveying" *Surveying World* May/Jun 1997:27-29.

Rosa, R. (1998) "National Mapping Bureau Airport Survey for the Office of Civil Aviation" *33rd Survey Congress, Association of Surveyors of Papua New Guinea,* Madang, July 1998.

Schupler, B.R. and Clark,T.A.(1991) "How different antennas affect the GPS observable" *GPS World* Nov/Dec 1991: 32-36.

Sipuman, J. (1998) "Shallow Water Mapping Applications using Satellite Remote Sensing (SPOT) at Bipi Island Manus Province" *33rd Survey Congress, Association of Surveyors of Papua New Guinea,* Madang, July 1998.

Spilkler, J.J. (1986) "Signal Structure and Performance Characteristics" *Global Positioning Systems* (Vol II), The Institute of Navigation: London.

Stoltz, A. and Morgan,P.J.(1989) "Monitoring Crustal Motion in Papua New Guinea using the Global Positioning System" *24th Survey Congress, Association of Surveyors of Papua New Guinea,* Port Moresby, June 1989.

Tolkki, S. and Koskelo,I.(1993) "Seeing the trees through the Forest" *GPS World* Oct 1993:21-27.

Woolard, D. (1997). "Survey and Mapping needs real-time DGPS" *Surveying World,* May/Jun 1997: 41-42

7

Inservice Environmental Education for Primary School Teachers: The Key to Environmental Education in Papua New Guinea

F. L. (Rick) Bein and Peter Clark

ABSTRACT

Environmental awareness in Papua New Guinea is the key to the conservation of pristine habitats throughout the country. School teachers are well situated in their communities to be major players in this effort. Environmental education can be carried out most effectively though the community and primary schools to reach large numbers of Papua New Guineans in the shortest amount of time. An environmental studies programme launched by the University of Technology Environmental Research and Management Centre and the Rainforest Habitat with the cooperation of the Morobe Department of Education and Wildlife and the Conservation Society of the Bronx Zoo began in 1998. Funded by K60,000 from the New Zealand High Commission, 100 grade school teachers from Morobe Province came together at the University of Technology in April to complete a one-week workshop on environmental studies. The Wildlife Conservation Society provided personnel and methods and materials already proven in other parts of the world which are now being tested for appropriateness in Papua New Guinea. Teachers are using the materials throughout Morobe Province and sending in periodic success reports. A basic "Introduction to the Fauna of Papua New Guinea" was also produced and distributed to all teachers involved to link the study material more specifically to Papua New Guinea.

Environmental Education Workshops for Teachers: The Mission and Philosophy

Papua New Guinea is preparing for the 21st century during which there will be major global pressures placed on its resources and its environment. Teachers have an important role in preparing Papua New Guineans for this event. The role in informing their students through environmental studies of the issues of environmental conservation, ecological balances, sustainable development is of paramount importance to the future quality of life in Papua New Guinea. With two-fifths of the world's original tropical rainforests already destroyed (Miller 1996), more attention has been given in recent years to Papua New Guinea's extensive areas of untouched timber resources. But most village people are in a poor position, educationally, to consider all the ecological ramifications of what the removal of the tropical rainforest could mean to their local environment. Whilst some exploitation of their natural resources must occur to achieve a higher standard of living, environmental education at an early age should at least provide some knowledge to help make these future critical decisions while balancing this very tricky equation.

In place of clear cutting of forests, selective logging by companies and by village operated *Wokabaut* sawmills can greatly reduce ecological damage to the rainforest if strict guide lines are followed. Because government control at project sites is meagre, it will increasingly become the responsibility of the local land-owners to help police the harvest of their forests. But the local landowners need some basic environmental knowledge to help them monitor logging companies and use *Wokabaut* sawmills wisely and ensure minimal damage to their environment.

Environmental studies is an interdisciplinary subject which lends itself to the teaching of many school subjects including not only science, geography, social studies, and agriculture, but also providing applications in the teaching of maths, English, art, music, home economics and history. We argue that the excitement and mission associated with environmental studies provides a stimulus to teaching the other subjects, and students so involved in their studies will also score better in their overall school performance. Teachers are in a very critical position in Papua New Guinea to bring about a significant environmental awareness among their people. Their position is strengthened by their influence on their students and as leaders in their communities. Students are a captive audience and must participate. Teachers can be further empowered with the enthusiasm and mission associated with conserving Papua New Guinea's natural environment. They can make a significant long term difference.

The environmental studies workshops were specifically designed to support community and primary schools in their efforts to teach environmental studies. It is our belief that a greater impact can be achieved at these junior levels. Young students are more impressionable because of their youth and especially when animals are used as a medium for teaching environmental studies (Berkovits 1994). As children learn more about the animal, they learn more about where the animal lives and what it requires for survival. Participatory learning including the use of animals, hands on activities, field excursions and real manipulation of the physical environment builds on what the children have already experienced and brings on a new perspective that engenders a longlasting interest and enthusiasm when introduced at younger ages. Papua New Guinea is just initiating an environmental studies programme in the elementary schools (Konaka 1997). We agree with this move to introduce environmental studies at this early age since most children in Papua New Guinea never proceed beyond the 6th grade and would not otherwise be exposed to environmental studies. Most of these younger children are the ones who will stay at home and remain to manage their local environments when they become adults. It is in the local communities that environmental thinking must be established. The National Environmental Education Advancement Project (NEEAP) states:

> Although environmental literacy is the primary goal of environmental education, it is not the end-all. Developing responsible environmental behaviour and improved environmental quality is the goal. Today's students are tomorrow's leaders and decision-makers. They need to learn and practise the skills necessary to protect, preserve, and restore the environment. Quality environmental education will lead to the acquisition of knowledge, the development of analytical skills, the beginning of environmentally conscious attitudes, and ultimately, to responsible behaviour (National Environment Education Project 1997).

Encouraging children to work with wildlife at an early age can have profound effects on attitudes toward the whole natural environment of Papua New Guinea. Such knowledge will remain with them and will have an impact on whatever they do in later life.

Description of the Environmental Workshop

The Rainforest Habitat and the Environmental Research and Management Centre (ERMC) arranged and organised environmental studies workshops for two groups of primary and community school teachers in early April 1998. The New Zealand High Commission and the Bronx Zoo of the Wildlife Conservation Society of New York jointly funded the workshops. Peter and Diane Clark of the Rainforest Habitat and Rick Bein of the Unitech ERMC were the principal collaborators in arranging the workshops. Morobe Province was selected as a test case to show how environmental education can enrich the curriculum in language, maths, art, social studies, and science (Bronx Zoo Education Department 1995). With the collaboration of the Morobe Department of Education, the National School Inspectors selected 52 schools from all over Morobe Province to participate in this landmark effort to infuse environmental consciousness into the schooling process. Teachers came from as far away as Kabwum, Siassi, Morobe, Wau and Kaiapit to participate in the five-day intensive workshops. One fourth grade teacher and one sixth grade teacher represented each school. In all 100 teachers attended the two workshops. Most teachers completed the environmental studies workshop full of enthusiasm and eagerness to start enriching their teaching with many new techniques associated with the hands on methodology.

The Bronx Zoo sent three environmental educators, Tom Naiman, Anne Robinson, and Alison Ormsby to conduct the workshops and the Zoo donated a large environmental teaching package containing video tapes, lesson plans, and study programme units for each teacher. 250,000 pages of text were reproduced for the 4,000 participating schoolchildren by the University of Technology.

Follow-up Activities

Follow-up activities and reports by the teachers are continuing as teachers

1. take a leadership role in their schools in setting up programs for World Environment Day;
2. bring their students on field trips to the Rainforest Habitat; and
3. complete their report forms which are then forwarded to the Bronx Zoo.

Already participating students are greeting visitors to their villages with comments on environmental issues. Ongoing communications from the teachers report on how well their programme is progressing and how much the experience has meant for their teaching methodology and their overall teaching mission. The participant teachers have requested more workshops and teaching materials. More than twenty-four teachers, on their own initiative, are presenting their own inservice environmental education programmes for their colleagues and for neighbouring schools. With the help of high school students and their instructor at Coronation High School in Lae, an environmental newsletter for the workshop teachers has been established. The purpose of this letter is to maintain continuity with the teachers and to encourage the continuation of the enthusiasm produced by the workshop. The first newsletter was mailed by the end of May 1998 to provide ideas and encouragement for the 5 June World Environmental Day. A second issue went out in August which reported on the events of that Day and provided encouraging ideas for teaching about the environment.

Conclusion

Going by the results of the workshops presented we would say that the programme was a major success. The workshops came off exactly as planned, including the arrival of 111 very interested teachers. The phenomenal teacher enthusiasm (thanks to the expertise of the Bronx Zoo demonstrators) during and following the workshops was tremendously encouraging. This enthusiasm is derived from the following factors.

1. Environmental studies is a relatively new subject in Papua New Guinea schools and little curriculum material has been developed so far.
2. The subject itself is very interesting because of the presence of the natural outdoor environmental laboratory available at most of the schools.

3. It is an easy topic to sell to students and to obtain the children's enthusiasm with the hands on activities, dramas, menus, personal experiences and local fauna.
4. The teachers' enthusiasm for the subject is contagiously mirrored by the students.

Final results of the workshops will not be known for many years until the children grow to maturity and begin taking responsibility for their own environments. Ongoing information gathered by the participating teachers, schools and communities will help in the assessment of this project's overall success. Although the current lack of information regarding the present levels of environmental damage that already exist in these rural areas will certainly make future analysis difficult, the real aims of the project are to effect a general attitudinal change in rural and urban village communities.

Although many Papua New Guineans live in nearly pristine environments, they are still somewhat distant from the understanding of how its different parts function and intertwine to make up the total ecosystem. At the local level in PNG, people know a lot about the fauna and many of the individual parts of their environment and their uses for food, medicine, building materials and clothing. But what is needed is the understanding that all of these parts rely on each other and only together do they make the environment function properly. If all goes well, there will be a repeat performance of the environmental inservice training courses for another 100 teachers in 1999. Much depends on the continued results of the programme assessment. Three other participating agencies must continue to endorse the environmental workshop. These are the Morobe Education Department, the New Zealand High Commission and the Wildlife Conservation Society.

REFERENCES

Berkovits, Annette and others (1994) "Bronx Zoo/Wildlife Conservation Park School Services". New York: The Wildlife Conservation Society Education Department.

Bronx Zoo Education Department (1995) *Pablo Python Looks at Animals*. New York: New York Zoological Society.

Konaka, Rachael S. (1997) "Primary Schools Introduce Environmental Studies." *Diwai* 2(11): 1-2.

Miller, G. Tyler Jr.(1996) *Living in the Environment* (9th ed). Belmont, Calif.: Wadsworth Publishing Company.

National Environmental Education Advancement Project (1997) *E2: Environment & Education, Creators of Environmental ACTION*. Boulder, Co.: NEEAP.

8

The Environmental Impact Statement in Papua New Guinea

F.L (Rick) Bein and John Rivers

ABSTRACT

This chapter discusses the role of the environmental impact statement. The view is taken that training students for environmental impact reporting in Papua New Guinea is best done with a broad multi-focus programme. Environmental studies is an applied science with an important and expanding role to play in the public policy and decision-making processes. With its practical orientation and its emphasis on place it is well placed to stimulate Papua New Guinea based research and to help build Papua New Guinea science studies at the University of Technology.

Introduction

The environmental impact statement is a formal written presentation prepared by a resource developer for assessment by a government authority and all affected parties that describes the natural environment in its current state and what the natural environment might become if the proposed development project is permitted to proceed. In Papua New Guinea the environmental impact statement is called an Environmental Plan (EP). The purpose of the EP is to alert all affected parties of all the impacts, negative and positive, of the proposed development project and to allow, in the event that the negative impacts are seen to be overly severe and not adequately compensated by the positive impacts, for the project to be stopped before any environmental or social impact has occurred.

The practice of making environmental impact statements began during the 1970s in the United States, Canada, and Western Europe (Miller, 1996). The requirement for this kind of reporting arose after decades of repetitive environmentally destructive industrial developments and the gradual rise of an environmental ethic and an increasing awareness that development does not necessarily need to sacrifice the environment.

In Papua New Guinea the sacrifice of the environment mostly supports the extraction of timber and minerals. This development takes place mainly to the detriment of Papua New Guinea's forests. One of the last remaining high bio-diversity tropical rainforest environments of the planet is at stake. Environmental degradation is occurring and with the current economic crises it will probably increase. While the economic benefits of the exploitation of the environment are immediate and obvious the dis-benefits to the environment are not always readily realised. It is difficult for the people to appreciate the problem until they have experienced the degradation first hand. By then it may be too late.

In Papua New Guinea project developers are required by law to prepare EPs for all proposed major development projects. Often perceived as a "necessary evil" the EP is usually assembled quickly and with minimum expense and with varying degrees of integrity. Some EPs are copied from previous plans with the names and dates changed. Site inspections may or may not occur. Planners may be constrained to look only at the immediate impacts of the development within a narrow timeframe and there may be no attempt to look at the broader picture. Often individuals preparing the EP have too narrow a technical or scientific background to enable them to adequately address all the ramifications of the project that they are reporting on. Typically, the foresters and miners who make the plans, while possessing a more than adequate background in the technical aspect of their industries, may have only very limited expertise in areas such as the historical, geographical, socio-economic, cultural and ecological aspects of the projects they are assessing.

After completion by the developer the EP is presented to the Papua New Guinea Department of Environment and Conservation (DEC) for review. Copies of the Plan are also given to the local communities and landholders who are affected, to local government and to relevant national institutes such as Forest Research Institute and the Department of Mining and Petroleum. Often the Plans are given little scrutiny except by the DEC officials who are required to formally accept the plan with a signature before development is allowed to proceed. However, DEC officials work under great difficulties, experiencing heavy workloads, depressed salaries, much political pressure and low morale that when combined may mean that the environmental assessment process is sometimes not fully completed and an inappropriate EP is given approval.

Problems of Legal Policy in Papua New Guinea

The primary piece of legislation governing the environment in Papua New Guinea is *The Environmental Planning Act* (EP Act) passed in 1982. The passing of the EP Act followed an international move in the early 1980s for governments to make legislation at the national level governing the relationship between development and the environment and setting standards for the protection of the environment. In these respects the EP Act followed "best practice" as it was understood at the time and contains clauses reflecting recent advances in environmental thinking such as:

> "environment" means the total stock of physical, biological and social resources available to man and other species and the ecosystems of which they are a part; (EP Act I, 2).

The EP Act widened previous legal definitions of "environment" to include not only flora and fauna but also natural features and "social resources". The EP Act was intended to be the primary vehicle for the assessment of environmental impact of major development projects and requires developers to submit for approval environmental plans and impact assessments. In these respects the EP Act is unexceptional. In practice, however, the EP Act has proved to be "dead law" in the sense that since its passing it has engendered no prosecution, nor is it likely to do so. The reasons for this are complex. They include: the exceptionally wide powers of ministerial discretion that are granted under the Act; the assertion within the Act that matters of national interest will overrule environmental protection conditions; the limitation of the juridical scope of the Act (which had the unintended and undesirable consequence of forcing the Lower Ok Tedi landowners into an offshore jurisdiction); and, possibly, some deficiencies in the administration of the Act.

What is of interest in this context, however, is not so much the limitations of the Act and its administration, but rather the underlying policy problems. The principal problem is that while the Act more or less satisfactorily performs its allotted task at the national level, it is now widely recognised that the protection of the environment must, in order to be effective, be grounded in local interests. The role of national legislation is to set certain standards, to ensure uniform codes, to establish principles, and to supply a legal framework for local activisms. Accordingly, many countries (e.g. UK, Australia, New Zealand) have tended in recent decades to make legislation the national interest and regulation the local interest. The duty of care for the environment is found in its most effective form at the local level and an important purpose of national legislation is, therefore, to give an effective legal voice to local and community interests. It is this latter feature that is absent in Papua New Guinea.

The Structure of Law in Papua New Guinea

The structure of law in Papua New Guinea is unusually complex. *The Constitution*, while establishing the primacy of parliamentary legislation also recognises the existence of custom law and common law. The relationship between these three kinds of law is not fully worked out in *The Constitution* and it has been neither clarified by later legislation, nor by judicial decisions. Among the many difficulties is the fact that *The Constitution* does not identify "customary law", but rather talks only of "underlying law". It was the intention of the authors of *The Constitution* that subsequent Acts develop the concept of underlying law by defining and incorporating custom law points and principles into the legal system. This has not been done. Nevertheless a hierarchy of laws has been established (*The Constitution* s9 and sch. 2.1). Following Orr on this point:

> Customary law is subject to written laws and common law is subject to both.
> Firstly, customary law is not adopted where and to the extent that it is inconsistent with a Constitutional Law (*The Constitution* and the Organic Laws). There may in particular be clashes between, on the one hand custom, and on the other hand the National Goals and Directive Principles, the Basic Social Obligations and the equality of citizens (s550). In all cases the constitutional requirements prevail.
> Secondly, "custom" is subject to statutes ... [and under the powers granted by the *Organic Law on Provincial Government* (OLPG)] ... the provincial Act, regulation or rule will prevail over custom (Orr 1991:78).

The effect that customary law has on common law has not yet been established, nevertheless it appears that where there is an inconsistency, custom applies and not the common law (Orr 1991:79). It is unlikely that any significant advances will be made towards clarifying this situation for a considerable period of time.

The existence of these problems within the law has, perhaps, had the unfortunate effect of diverting attention away from the underlying policy problems. Whatever the legislative confusion there remain the larger problems of the relationship between law and civil society.

Local communities, with a solidarity defined through landownership and perceiving the state to have failed in its developmental tasks, have in recent years frequently confronted the state with assertive sovereignty and compensation claims. The response of the state has been, in many cases, to avoid confrontations and to meet local group compensation demands for both real and imagined wrongs. The moral basis of these claims is to be found in the custom law of the group. But the same landowner group, pursuing similar custom law claims against a near neighbour may find itself enmeshed in a violent conflict. In these contemporary regional conflicts the police and the state are reduced to being largely powerless bystanders playing, at best, a mediating role between the warring parties.

What is absent is a regional civil society that transcends the moral community of the village. What is absent is the structure of Act, regulation and rule that both support and is supported by the institutions of a regional civil society. We refer here to those aspects of law and society that have little or nothing to do with policing but nevertheless play an essential role in supporting the politics and economics of daily life of all citizens. They have the function of regulating and supporting markets, exchanges, rights and obligations, entitlements, of being the rules of everyday life. These matters are part of every developed society and their role in the growth and development process has gained in recognition in recent decades.[1]

Law and Development in Papua New Guinea

Since its enactment in 1978 the *Organic Law on Provincial Government* has been legislatively revisited many times. The discussions have usually focussed on the question of giving more autonomy to the regions to arm them against the perceived overweening power of the centralised state. Politically this has been seen as a contest of sovereignty between the local and the central. This, we think, is a mistake. Provincial government in Papua New Guinea does not effectively represent any local environmental principle, nor can it be reasonably expected to do so. It is simply too distant from the myriad of local communities that make up each region. Provincial government can expect no better success than national government in this respect.[2] Nevertheless the changes to the OLPG have had the effect of creating the possibility of locally grounded regulation in some areas, particularly in relation to the environment and in the cities. Recent discussions with officials on these matters have led to complaints being aired on the turpitude of State Government on many legal matters. Our response has been that to wait for the state is to wait forever and in any case the state cannot address these problems. What is called for are positive initiatives on the part of responsible officials that will have the effect of creating the conditions that inspire and support the development of appropriate Acts and rules and regulations at the regional and local level. In short the role of the administrator in stimulating legal advances through generating effective rules and regulations at the local level should not be overlooked and, given the development and environmental conservation needs of Papua New Guinea and the necessity to build and strengthen the bonds of civil society, a society that transcends the village, these matters should be given some priority.

When a mineral or forestry resource extraction company enters a region for the first time it typically deals with landowners, villagers, forest dwellers, who have been waiting for the state to bring development to them for a very long time. They have expectations for the modern. Yet when the company officials step out of their offices and into the village or into the forest clearing to meet with landowners they step into a custom law setting. When Environmental officers, representatives of national and local government, consult with landowners they hear "talk" which comes from a custom law context and insofar as the talk raises disputatious matters they are so in reference to the villagers' aspirations for justice to be done according to custom. Land is the physical basis of the sovereignty of the community and custom law is the cultural basis of the sovereignty of the community. In order to deal with these matters effectively DEC officers, Provincial Government officers and local level government officers, Environmental officers, charged with the duty of care for the environment must have a knowledge of, and sympathy for, custom law issues.

When they are dealing with development proposals and community development issues Environmental officers are in a position, through the exercise of a knowing judgement, to lay the basis for future institutional development. At the beginning of, say, a mine project the host landowner community may be viewed as traditional. Before ten years have elapsed this community will have been transformed by the process of economic differentiation, migration in and out, employment and unemployment, the rise of new forms of politics, new kinds of households, all the factors that accompany development and when combined create a sharp sense of difference between people where none existed before. Traditional forms of dissensus frequently continue and may also be sharpened by the movement to the modern. Traditional forms of consensus are found inadequate. New forms are required.

There are three main requirements that must be met for endogenous economic growth to take place in a developing country. The three requirements may be expressed as "gaps" (Barrett 1997). The first (and best understood) is the "object gap" — capital, human capital and embodied capital. The second is the "idea gap" — the transfer of ideas from the developed world to the developing country. The third is the "trust gap" — in a developing country there is a relative absence of the set of shared expectations, norms, rules and institutions that support the interpersonal exchanges that lie at the heart of the economic life of the market, town and city in a developed society. This is the social element in economic development. It is easy to underestimate the extent to which the institutions of a developed society support modern forms of interpersonal exchange because their ubiquity ensures that they form mostly a backdrop to people's conscious lives. The formal legal system of a country constitutes the projecting tip of this complex of ideas and expectations. But the legal order would not exist without a supporting framework of social practices and attitudes that simultaneously support and constitute the element of trust that is part of all social and economic exchanges and define where trust ends and bad dealing, malfeasance, misfeasance and nonfeasance, begins. The development of this social and legal structure provides the basis for the modern concept of common property and common property rights on which modern environmental protection legislation depends.

Development economists have devoted considerable efforts towards the definition and analysis of the object gap, the need for capital and embodied capital, in the developing countries. In recent years some attention has been paid to the ideas gap, both to limit and to aid the transfer of intellectual property to developing countries. In each case there is an obvious and prominent role for North South transfers in overcoming these gaps. But overcoming the "trust gap" in a developing country is strictly a local problem. The gap may only be filled and the institutions created by the small and incremental gestures of individuals. Institutions cannot be imported into the country on consignment like rice and computers. There is no Asian Development Bank loan for this purpose. It is our view that it is in precisely this area that Environmental officers can fulfil a most important role. Working at the critical interface between custom law and state law they are able to supply a vision of trust and through their knowing use of rule and regulation establish a modern context for the development of the common property rights which support a broad based community support for the care of the environment.

Changing Views of the Environment

In the traditional setting of Papua New Guinea no direct thought exists towards preserving the natural environment for its own sake. The natural environment is viewed as the all providing "mother" upon which no demand can be too great.[3] This view that the "environment can take care of itself" has sustainability as long as the resource use does not exceed nature's limits to support a certain number of people at a given standard of living (Hardin, 1968). If this is reality, a balance with nature can be maintained indefinitely.

The reality in contemporary Papua New Guinea, however, is that resource use is exceeding nature's limits to provide. Growing human populations (Papua New Guinea is doubling its population every 30 years) and invading western living standards, including the monetary economy, exert ever-greater pressures on the environment. Yet, the traditional thinking that "nature will provide" still prevails in Papua New Guinea. Many developed countries have gone through their development transformations while continuing to think in this way. It is in this stage of their economic development that most nations have drastically altered their natural environments and before they realise what they have done they have passed beyond a "point of no return" (Miller 1996). It is most precisely at this point in time that a concerted educational effort must be made to alert the people to the future consequences for the environment of present decisions and the possibility of irreversible future damage.

This is a critical time for Papua New Guineans. Economic development is foremost in the minds of decision-makers rather than concern for the environment. A serious environmental awareness effort is necessary in Papua New Guinea to forestall the severe destruction that has already overwhelmed other island nations such as Madagascar, Haiti, Jamaica, and the Philippines. Such an effort must be directed to bring Papua New Guineans to the realisation that they need to actively participate in the guardianship of their environment.

Education and the Environment

The preparation, interpretation and follow-up of environmental plans are an excellent place to begin this participation. This is a large task which can involve education at all levels. The two major universities have been programmatically empowered to provide a broadbased education and to take the leadership in this effort. Both institutions, however, were conceived and set up before environmental thinking, environmental activism and environmental stewardship were part of the agenda. Therefore, the universities must change their thinking, priorities and programmes. The wide variety of Academic Departments, Centres and Programmes which have environmental science and environmental impact evaluation components can provide the institutional and knowledge bases for the task of teaching about the functions and the care of the environment. Environmental studies occupies a pivotal position in science and technology in Papua New Guinea and can serve as an essential coordination point and provide the needed Papua New Guinea studies focus for contributing scientific, technical and professional disciplines.

That Papua New Guinea should avoid this environmental point of no return is also a major concern for the global environment. Many of the developed countries are interested in the protection of Papua New Guinea's bio-diversity and are ready to provide financial assistance if there was a reasonable expectation that money would be used for this purpose. The global interest has become more sharply defined in recent years. Presently Papua New Guinea's contribution to greenhouse gas emissions is being surveyed. Tradable rights to greenhouse gas emissions have undergone a rapid development in the last two years. We propose that a portion of the monies that Papua New Guinea will receive from this scheme be devoted to University environmental education and the training of graduates with skills in the area of environmental impact assessment. There is a growing need for these skills at all levels of government, NGO and community activism and in the project development industries.

NOTES

1. See for example: Boeninger (1992); Larmour (1997); Martin 1992.

2. Letter writers to the *Post Courier* newspaper frequently assert that Provincial Governments simply reproduce the kleptocracy and neo-patrimonialism of national government. *Post Courier* 7/5/98, 11/5/98, 14/5/98.

3. Exceptions may occur inadvertently when natural areas are set aside for whatever reason to appease the ancestors or other forms in the spirit world.

REFERENCES

Barrett, C.B. (1997). Idea Gaps, Object Gaps and Trust Gaps in Economic Development. *Journal of Developing Areas* 31: 553-568.

Boeninger, E. (1992). "Governance and Development: issues and constraints," *Proceedings of the World Bank Annual Conference on Development Economics - 1991*. Washington: World Bank.

Hardin, G. (1968). "The Tragedy of the Commons" *Science*, 162, 1243-48.

Larmour, P. (1997). "Corruption and Governance in the South Pacific" *State, Society and Governance in Melanesia*. Discussion Paper 97/5, Research School of Pacific and Asian Studies. Canberra: Australian National University.

Martin, D-C. (1992). "The Cultural dimensions of Governance" *Proceedings of the World Bank Annual Conference on Development Economics – 1991*. Washington: World Bank.

Miller, G.T. 1996). *Living in the Environment* (9th ed). Belmont: Wadsworth.

Orr, R. (1991). "Provincial Government and Customary Law" *Melanesian Law Journal*, special issue, 79.

9

Sustainable Development and the Environment in Papua New Guinea

John Rivers

ABSTRACT

Although Papua New Guinea is a developing country there has been little investment in development education. Development studies is not part of the secondary school curriculum and has little visible presence in the universities. This chapter begins by giving a brief account of why development studies was omitted from the foundation curricula of the Papua New Guinea universities, and continues by arguing that in the past decade development studies has advanced to the point where it can make practical contributions to sustainable development in Papua New Guinea and to meeting the growing demand for trained personnel to work in the field of environmental conservation and sustainable development at the village level.

Introduction

Although Papua New Guinea is a developing country there has been little effort devoted to the teaching of development studies within the country over the last few decades. The reasons why development studies was not made an integral part of the higher education system in Papua New Guinea are partly historical and partly political. At the time when the University of Papua New Guinea and the Papua New Guinea University of Technology were established development studies was an intensely politicised discipline making only a slight claim to useful technical knowledge. The main focus of interest in development studies during this period was on the critique of colonialism and neo-colonialism. Accordingly, although the Papua New Guinea universities were intended to serve national development goals, development studies was not seen as having the potential to make a practical contribution towards the realisation of these goals in the university context and was not included among the foundation faculties.

Since that time, and especially in the last decade, there have been some significant changes in the discipline of development studies resulting from advances in the study of economic growth processes, project design and evaluation techniques, development administration and policy studies. During this same period there has been a corresponding decline of interest in the 'isms of Marxism, socialism and dependency theory. Because of these changes it is now timely to re-consider the potential contribution of contemporary development studies to sustainable development in Papua New Guinea.

History[1]

The origins of development studies as a discipline date from the 1930s and the technical discovery of malnutrition in India. Following from these discoveries there arose a technical and policy interest in the welfare of the populations of the colonial states. This interest became of paramount concern in the formative years of the United Nations where there was a concerted attempt to find technical solutions to the problem of achieving growth in the colonies as they progressed towards independence after World War II. From the establishment of the International Bank for Reconstruction and Development (now the World Bank) in 1948 and for the next two decades development studies was dominated by economists. During this period many of the problems of capital transfer, capital investment and aid flows between the developed countries and the less-developed countries were widely discussed. The various attempts to create economic growth during this period were not, however, notably successful.

Within the World Bank the lack of success in achieving rapid progress in implementing economic development in the third world led to the establishment of the International Development Association (IDA), the soft loan arm of the World Bank. Outside the World Bank there was a burgeoning interest in the academic study of development and the establishment of the study of development problems in the universities. During this period we see the rise of alternative theories of development which fused sociology and economics and developed a critique which argued that the primary cause of the less developed countries backwardness was the first world's domination of the world economy. The essence of this domination lay in a system of structured inequality which maintained the less developed countries' in a permanent state of dependency, or dependent development, and which allowed their continued economic exploitation by the developed countries. The theory of dependent development, which originated in Latin America, and similar critiques of neo-colonialism gained wide currency in academic circles in the developed countries and in one or another version came to be widely believed by a very large number of people in the developing world. However by the mid 1980s academic interest in dependency theory had waned considerably, mainly because many of its core propositions came to be seen as having merely rhetorical force, rather than being based on well-founded fact and analysis.

Contemporary Approaches to Development

Accordingly, and for complex reasons associated with the deregulation of world markets (particularly the financial markets), the idea of a world system, able to be manipulated by nations or multinationals, lost favour and the focus of development studies changed course accordingly. To some extent this change was accelerated by contemporary changes in economic theory. A major influence was the revival of interest in institutional economics. Since the 1940s economics had been dominated by the neo-classical paradigm of a micro-economics which assumes that the foundations of economics are to be found in the rational choices of individual consumers. Macroeconomics is defined as the study of the aggregate of these choices. Neo-classical economic theory has some severe limitations when applied to the study of growth and development. Institutional economics concentrates on the study of the contribution of social institutions to the creation of markets and economic growth. Institutional economics has two main foci of interest. One of these is the analysis of transaction costs in an economy and the process of the growth and development of those efficient market institutions which lead to the lowering of transaction costs and the improvement in the efficiency of an economy. The other is the role of social institutions in creating, or in some cases destroying, those elements of trust that are necessary to support many types of economic transaction. These institutional aspects of economic growth and development are of great interest to those involved in analysing contemporary problems of economic development and planning for economic growth at both the national and the "grass-roots" levels in developing countries.

There are three main requirements that must be met for endogenous economic growth to take place in a developing country. The three requirements may be expressed as "gaps". The first (and best understood) is the "object gap" — capital, human capital and embodied capital. The second is the "idea gap" — the transfer of ideas from the developed world to the developing country. The third is the "trust gap" — in a developing country there is a relative absence of the set of shared expectations, norms, rules and institutions that support the interpersonal exchanges that lie at the heart of the economic life of the market, town and city in a developed society (Barrett 1997). This is the social element in economic development. It is easy to underestimate the extent to which the institutions of a developed society support modern forms of interpersonal exchange because their ubiquity ensures that they form mostly a backdrop to people's conscious lives. The formal legal system of a country constitutes the projecting tip of this complex of ideas and expectations. But the legal order would not exist without a supporting framework of social practices and attitudes that simultaneously support and constitute the element of trust that is part of all social and economic exchanges and define where trust ends and bad dealing — malfeasance, misfeasance and nonfeasance — begins.

Development economists have devoted considerable efforts towards the definition and analysis of the object gap, the need for capital and embodied capital, in the developing countries. In recent years some attention has been paid to the ideas gap, both to limit and to aid the transfer of intellectual property to developing countries. In each case there is an obvious and prominent role for North-South transfers in overcoming these gaps. But overcoming the "trust gap" in a developing country is strictly a local problem. The gap may only be filled and the institutions created by the small and incremental gestures of individuals. Institutions cannot be imported into the country on consignment like rice and computers. There is no Asian Development Bank loan for this purpose. It is my view that it is in precisely this area that development studies can fulfil a most important role in contributing to sustainable development in Papua New Guinea.

Academic interest in the political critique of development began to decline in the mid-1980s. The Berg Report's 1984 defence of World Bank Structural Adjustment Programmes may, perhaps, be credited with stimulating the current interest in the role of governance in economic development and the study of development administration. Further support for this change of focus has come from two additional sources. One of these has been the advances in the understanding of public policy issues and consequential improvements in the technique of project design, implementation and management. The other has been a revival of interest, spurred on by NGO involvement, in practical techniques for "grassroots" development, basic needs service delivery systems, sustainable development projects and community-based resource management. Implementing sustainable development and environmental management projects requires a detailed understanding of the sociology, economics and politics of the target group and a sophisticated grasp of the administrative techniques. In the 1980s a number of universities reflected these changes of interest by re-organising development studies and installing development administration and environmental resource management as core subjects within the development studies programme.

Achieving Sustainable Development in Papua New Guinea

Sustainable development in the Papua New Guinea context has some aspects that are common to other developing countries and some that are more or less peculiar to Papua New Guinea.

Landownership

Papua New Guinea is unusual in that a very large proportion of the land remains under customary ownership and in the control of the landowning group. Land ownership and inheritance are frequently phrased in terms of an ideology of descent, but in practice, control over land and access to resources usually depends on co-residence and residential proximity. Although, in many parts of Papua New Guinea the environment has been historically transformed by horticultural practices, even in these areas and in many other parts of the country, many people live in and adjacent to rain forests and other wilderness areas. Practically speaking, apart from the very small amount (possibly 3 per cent) of alienated land, all remaining land is under custom ownership, control and at least occasional economic use. The conventional first world concept of the "wilderness area" does not really apply in Papua New Guinea because much of this land is used to meet subsistence needs and even when not so used will be made the subject of a customary land ownership claim when outsider interests intervene.

Household Reproduction and Sustainability

Traditional households in Papua New Guinea met their subsistence needs from horticulture, fishing, hunting and gathering activities. The Papua New Guinea Household Survey of 1996 covered a random sample of 1,200 rural and urban households (Gibson 1998). The Household Survey provides graphic evidence of the continuing dependence of the rural and cash poor households on traditional subsistence activities. The consumption items that are most heavily skewed towards the poorer households are self-produced, directly derived from the interaction between the landowner and his family and their local environment. For the majority of households in Papua New Guinea environmental management and sustainability are not abstract questions but, rather, questions that are suffused with practical everyday meaning and with direct connections to the basic needs of the household and its subsistence activities now and into the future. The Household Survey found that "the items whose consumption is most heavily concentrated on the poor are locally produced foods and firewood" (Gibson 1998:34). A selection of the highest ranked consumption items is given in Table 9.1:

Table 9.1
Distributional characteristics of selected items consumed in Papua New Guinea Households

	Rank
Sago	1
Taro and Chinese Taro	2
Firewood	3
Yams	4
Cassava	5
Sweet potato	6
Leaf tobacco (and mutrus)	7
Banana (cooking and sweet)	8
Sugar cane	9
Other greens, vegetables and nuts	10
Aibika	11
Coconut	12
Axes, bushknives and other tools	13
Bush meat and other unspecified meat	14
School fees	15
Children's clothing (new and used)	16
Pots, plates, cups and other utensils	17
Fresh fruit (excluding bananas)	18
Salt, pepper, spices, sauces	19
Rice	20
Tinned fish	21

The Papua New Guinea Household Survey examined 87 consumption items. Of the 15 top ranked items only 2 have to be purchased. These are in rank order, Axes, bushknives and other tools (13) and School fees (15) (Gibson 1998:38). Other items listed that may be derived wholly, or in part, from hunting and gathering activities are: Firewood; Other greens, vegetables and nuts; bush meat and other unspecified meat; and fresh fruit. The natural environment in Papua New Guinea constitutes a rich and accessible resource for a large number, but not all, of the poor. In this respect Papua New Guinea differs from some other developing countries where access to resources and options for the use of endowments may be severely constrained, indeed the chief constraint that the rural poor face (Carter and May 1999:16).

The relationship between the Papua New Guinean person and their environment is more direct than that of the great mass of people living in the urbanised first world. It is also, to a large extent, outside the direct reach of government and the grasp of sectoral policy makers. Accordingly, the question of achieving sustainability and development in Papua New Guinea requires multiple local solutions.

Legislating for Conservation

The State has made efforts to initiate and promote conservation and sustainable use of natural resources in Papua New Guinea. Besides establishing a Department of Environment and Conservation (DEC) (but at the time of writing reduced to office status), the State has been active in passing legislation for environmental conservation and sustainable development purposes. An example is the *Wildlife Management Areas* Act. This Act provides for a local landowning group to elect to place some of their land under a conservation regime, administered by themselves with the assistance and support of NGO and Department of Environment and Conservation officials. But the legislature in Papua New Guinea is very distant from the village and legal instruments have little or no referential context in the village. Legislation becomes vapid in the village, which exists in its own political universe, remote in space and time from the capital. What are the end results of this kind of legislation in the Papua New Guinea village and remote area context? The example I discuss here comes from the Kikori area of Gulf Province and refers to a scheme put in place as a result of initiatives taken by the United States based Chevron Oil Company, which holds prospecting rights in the region, the international NGO, WWF, and the Papua New Guinea Department of Environment and Conservation. The difficulties of implementing legalistic regimes in the village context and the lack of fit between the conservation management scheme and the realities of village level politics, with their deeply personalised character, are evident in the following letter. The letter was written by a student fieldworker who was assisting Chevron and the WWF implement the Wildlife Management Area and other similar village conservation and development schemes. Sisa Kini, the writer, is expressing her sense of frustration over the tangle of unintended consequences and inappropriate outcomes that often accompany the implementation of these kinds of schemes:

> I was actually going to reply yesterday, but had a lot of paper work to complete for an

awareness campaign we are going to carry out tomorrow on the other alternatives that are available [for the villagers] to large scale commercial logging. And also to make them aware of the importance of hearing how these projects affect the lives of everyone in the community, including the women so that they [will] allow the women to participate a lot more in community decisions.

The problem I wanted to ask for your input . . . was about a Wildlife Management Area (WMA), here in Kikori, called Ifa. Our desk officer at DEC in the WMA's & National Parks Section came over from Port Moresby on Monday (all expenses paid for by WWF) to find out what the situation is like with the Ifa WMA committee. Why? Because a letter was delivered to them without the consent of the Chairman they have gazetted the DEC officer wanted to find out by coming to the village what the situation was like and to see if these people had decided who the new committee members were. When he came he found that the two chairmen, the old & the new/contesting one, were both in Port Moresby. Therefore we couldn't get a response from them. But the rest of the committee members & various leaders, after a long discussion, soon revealed the situation.

The new committee is exactly the same as the old one. The request is only for a change in the chairman & another member, or for these two to leave the committee. The new and the old chairman went to court at the district court, but have been sent back to local level to mediate the issue. So this is the current situation. What I found whilst listening to them is that many of their complaints against the old chairman are as a result of him not giving them information and being inactive. So maybe he is inactive, because he too doesn't understand.

The questions they have asked us were put indirectly. I felt we didn't adequately satisfy them because they requested that we go back again, when the new committee is finally decided, and that this visit must have Chevron officers, WWF officers, & Government officers [present]. Possibly they feel that these three parties to the WMA project have things to clarify for them too. I really feel they don't understand what a WMA is? — the legislation surrounding it? — the benefits to be obtained from having one? — the rights and obligations they have etc. For instance they asked if "Ships/tug-boats using the waterways within the WMA, do they have to pay some sort of compensation and is this okay?" The DEC officer did not properly address this issue . . .

Also I noted there were no women present at all in this meeting except for myself and I'm not a member of the community. I feel it's important for all members of the community to be aware, but it seems that men make all the decisions. Maybe women have their say at home, and the men present these views. This is so as to avoid public conflict. Should this change? And why? Should the women have direct say? What difference would this make? These are some of the questions I am working on currently.

What is an effective response in this case? Bringing together and joining village politics and modern legal concepts to create an effective conservation instrumentality is a formidable task. There is, all too frequently, an underestimation of the length of time required to build the required institutional capacity. Modern institutions rely on institutionalised trust. Institutions are formed over decades, not months. To attempt to simply emplace them in the village context is to court failure and in many cases to mistake policy for process.

Improving Project Implementation

In Papua New Guinea in the 1990s, as in many developing countries, the role of international and national NGOs in implementing development and conservation projects and administering and extending basic needs services has expanded enormously. Increasingly, however, aid and NGO donor organisations have felt pressures to audit and justify their expenditures on development and conservation projects. For example, in the case of a basic needs and services project such as a village water supply, aid organisations are looking to achieve, firstly, economies of scale by concerting a large number of small projects together and, secondly, project sustainability by introducing project management systems that ensure that system maintenance requirements are met in the context of the village. Relief aid can be delivered out of the back door of an airborne Hercules. Sustainable village development and conservation projects cannot. Achieving the goal of project sustainability requires the use of trained personnel with a sophisticated understanding of local land tenure, kinship, and the social organisation of the targeted communities. There are numerous instances of failed development and conservation projects in Papua New Guinea (Rivers nd). There is now a general consensus that the standard of accountability used in development project work should be broadened to include appraisals of project effectiveness over the longer, rather than shorter, run. This requires project work to be done to a higher standard, in many cases using the same technology, but raising the standards of community assessment and appraisal and including a social institution capacity building component at the "grassroots" level. It is only by the use of trained personnel, with a good knowledge of development administration and of village society that these goals can be achieved and development agencies can avoid creating a third generation of failures.[2]

There is evidence of a rising demand for community development trained personnel among the donor aid agencies, NGOs and others involved in planning and implementing development projects. Advertised vacancies in the Papua New Guinea press for such people are now running at about 80 per year. There is a proposal currently before the Office of National Planning and Implementation to require all development project proposals to include a person to carry out pre-implementation feasibility studies, to port the project into the village, to assist villagers to organise their communities and resources to meet project maintenance requirements and to audit the functioning of completed projects. This proposal, if implemented, is likely to lead to an immediate demand for about 300 trained personnel. Many of them would receive employer sponsorship for their studies if a course were available. The willingness of employers to sponsor studies in this area is an indication of the extent to which demand exceeds supply. There is a huge untapped demand for development education in Papua New Guinea.

It is timely to recognise that the problems of long term environmental management and sustainable village development will require solutions that are considerably more sophisticated in terms of project design, implementation and accountability than have been applied to date. It is timely to recognise the extent to which these initiatives will require the services of personnel with a specialised training and a deep understanding of village culture. Caring for the environment in Papua New Guinea requires, firstly, that we understand the environment in the scientific sense. But this understanding will only become effective when we develop the capacity to put it to use to meet village people's development aspirations and create sustainable development at the village level.

NOTES

1. For an extended discussion of the history of development the reader is referred to Meier and Seers (1984) and Meier (1987).

2. The normal contract duration for an overseas volunteer is two years. A frequently cited "rule of thumb" in Papua New Guinea states that a foreigner working in the village context usually requires about eighteen months' contact to gain sufficient experience to begin to understand the intricacies of village kinship systems, politics, and landowner issues to the point where they are able to accurately assess the workability of project proposals at the village level.

REFERENCES

Barrett, C.B. (1997). "Idea Gaps, Object Gaps and Trust Gaps in Economic Development." *Journal of Developing Areas*, 31, 553-568.

Carter, M.R. and May, J. (1999). "Poverty, livelihood and class in rural South Africa." *World Development*, 27(1): 1-20.

Gibson, J. (1998). "Indirect tax reform and the poor in Papua New Guinea," *Pacific Economic Bulletin*, 13(2): 29-39.

Kini, Sisa. (1999). Personal communication.

Meier, G.M.(ed) (1987). *Pioneers in Development - Second Series*. New York: World Bank/Oxford University Press.

Meier, G.M. and Seers, D. (eds) (1984). *Pioneers in Development*. New York: World Bank/Oxford University Press.

Rivers, J. (nd). "A failed rice scheme." To appear in *Research in Melanesia*.

10

Knowledge of Traditional Agriculture in Training for Agricultural Research and Development

Alan R Quartermain

ABSTRACT

There is a wealth of information concerning the diversity of crop plants, traditional agricultural practices and farming systems of Papua New Guinea. However, there is much less knowledge or understanding concerning this diversity and how best to use it in rural development. Traditional knowledge or knowledge about traditional practices and systems could be of immense value in the search for sustainable intensification of production. Hence tertiary agricultural institutions should educate and train students to appreciate and understand such knowledge so that they may be enabled to further the understanding through research or utilise the knowledge for the benefit of farmers.

Introduction

Farmers or gardeners in Papua New Guinea producing food mainly for family subsistence have a wide range of technologies and systems which they use to achieve their ends. Insofar as these practices have roots going back into pre-colonial times they can be regarded as traditional and knowledge of their use or effectiveness is part of the heritage of traditional subsistence farmers. This is distinct from the knowledge of such practices that may or may not be possessed by educators and trainers involved in human resource development for agricultural research and development. The questions that are addressed in this paper are: How much is known about traditional practices by people other than the practitioners themselves? To what extent is such knowledge useful or likely to be useful to agricultural advisory services or in rural development? Can or should agricultural scientists and technicians be trained to research, understand and apply such knowledge? To what extent, therefore, should traditional agricultural knowledge be included in the curriculum or constituent subject syllabi used in relevant training institutions?

Discussion

The Knowledge Base

There is a bewildering diversity of food crops, agricultural practices and farming systems in Papua New Guinea. The reasons for this are well known and rooted both in the geography of the country and in its history of settlement and agriculture stretched over some 10,000 years. Traditional gardeners have readily invented, modified, adapted or adopted plants and technologies to suit their immediate purposes, where such have been needed or proven effective and beneficial. As a consequence the people have at best thrived and at worst survived.

There is now very good documentation of information on the food crops and agricultural systems used. Perhaps still the best general account of subsistence agriculture (here more or less synonymous with traditional agriculture) is that of Macewan (1978) while the unpublished (but available in University of Vudal and Department of Agriculture and Livestock Libraries) report of Kesavan (1983) contains a comprehensive overview of the systems and their diversity, crop cultivation methods and yields, changes occurring in the systems and the constraints to production. He includes four case studies, taken from the literature, covering the mound builders of Enga, mega-yam production in Maprik, the mid-altitude techniques of the Maring and sago utilisation in the Sepik.

Powell (1976) lists 227 plant species used as food, only 14 of which are noted as recently introduced. French (1986) lists 536 food plants in his descriptive compendium while in a more recent (1996) unpublished report on Research Planning for Food Crop Production in Papua New Guinea he lists 602 species made up of 29 starchy staples, 21 beans, 35 edible greens, 30 edible ferns, 67 vegetable crops, 138 fruit species (plus 43 species probably present in Western Province), 52 types of nut, 29 spices and 158 minor food plants. Of these he suggests 73 species for priority attention. French also notes that production of these crops is constrained by about 230 known disease organisms and approximately 600 insect pests.

It is also well known that there are enormous numbers of distinct cultivars or land races of the major staples. The approximate numbers would appear to be 5000 for sweet potato, 400-500 for yams, 300-400 for taro, 400-500 for bananas and over 30 for sago. There is also a rich genetic diversity for sugarcane, coastal and highland pitpit, rungia, aibika and winged bean. The most recent overview of plant genetic resources in Papua New Guinea is that of Kambuou (1996).

With respect to systems, the Papua New Guinea Agricultural Systems Project was designed to provide detailed information on agricultural practices and cropping patterns in small holder (subsistence) agriculture. Data were collected by field observation, interviews with villagers and reference to published and unpublished documents. The information is stored in coded form in a data base and has been mapped in a way compatible with the Papua New Guinea Resource Information System. A series of 19 working papers dated 1993 to 1997 has been produced covering all provinces except North Solomons and a compilation of text summaries (Bourke 1996). All this constitutes an incredibly detailed and valuable data base. The last published working paper, that for Morobe Province (Bourke et al, 1997), has been used to indicate the type of information included. An agricultural system is identified when sets of similar agricultural crops and practices occur within defined areas. Six criteria are used to distinguish one system from another:

1. Fallow Type (the vegetation cleared before cultivation).
2. Fallow Period (length of time the site is unused between cultivations).
3. Cultivation Intensity (number of consecutive crops planted before fallowing).
4. The staple or most important crops (including sago).
5. Garden and Crop Segregation (extent to which crops are planted in separate gardens, in separated areas within a garden, or planted sequentially).
6. Soil Fertility Maintenance Techniques (other than natural fallows).

Much more information on practices is included such as the placing of pigs in gardens, burning, soil tillage, deep holing, fencing, irrigation, mulching, drainage, soil retention barriers, staking, mounding and the making of various kinds of garden beds. Not all of the practices or optional techniques are traditional in a historical sense but most are now integral to the systems. While a precise count has not been made, there may be between 100 and 150 clearly definable, distinct systems. These systems can be related to any agro-climatic zoning system such as that of Gurnah (1992). All of this information however is simply information, not knowledge in a wider sense which implies understanding of purpose and, in a scientific sense, mode of action and effectiveness of the practices.

Traditional knowledge (sometimes called indigenous knowledge) is site specific and very personal, deriving as much from religious or mystical beliefs as from any understanding of agricultural processes. Why things are done or why they happen and why practices change over time are not always obvious to the outside observer, no matter how sensitive. Constraints to production are often social rather than technical. Religious belief and magical practice are often influential in determining, for example, what can be eaten, patterns of resource ownership and perceptions as to why agricultural practices are effective or not, what causes soil fertility, how plant disease may be prevented and why erosion occurs (Pernetta et al, 1982; Bulmer 1982). This personal knowledge cannot easily, if at all, be stored or made generally accessible.

Many practices, however, have obvious practical benefits regardless of the practitioners' perception as to how or why they work. A large number of apparently useful techniques can be found in the literature including soil retention devices, ditching and drainage, mulching, mounding, composing, irrigation, mixed cropping and fallowing (Pernetta et al, 1982, Wood and Humphreys 1982, Powell 1982, Spriggs 1982 - all in Morauta et al, 1982, Kesavan 1983). Traditional agricultural technologies can usually be seen to revolve around soil management, crop protection, maintenance of diversity, sowing and harvesting, the role of animals and selection of planting materials. All the practices have been developed, not with any longterm view of sustainability of production but because people have learnt from observation, inspiration, experience, discussion or trial and error that they are beneficial to immediate yields, food security and the welfare of the current land users.

The 1980 conference on Traditional Conservation in Papua New Guinea: Implications for Today (Morauta et al, 1982), which set out inter alia to examine the extent and nature of traditional conservation knowledge and practice and the relevance of these in Papua New Guinea today, somewhat clarified the scope, strengths and limitations of such knowledge with respect to the environment and environmental management. Some of the papers were directly concerned with agriculture, especially that of Wood and Humphreys (1982) in which they discuss traditional soil conservation methods for the control of soil erosion, fertility, water and temperature and that of Spriggs (1982) on the use of water in true irrigation, in wetland or swamp agriculture and in the pit cultivation of swamp taro (Cyrtosperma) on atolls. While much of the other material also has direct implications for agriculture, the agricultural research, teaching, development and production community would greatly benefit from the convening of a similar conference exploring what is known about traditional agriculture and a pulling together of the available information from the wide range of disparate sources.

The Usefulness of the Knowledge

As indicated in the previous section, there is a wealth of descriptive or classified information available on traditional practices, and in many cases on the practitioners' reasons for engagement. What is sadly lacking is in-depth analysis or research into why systems work, why some have periodically failed and why changes have taken place. Traditional management systems may have much to teach since there is knowledge that may be valuable to a modern society in search of viable models of sustainable development. This knowledge is also of scientific value but usually underrated by modern agricultural scientists and too little account is taken of it in the search for research projects that might yield quick returns on investment. Traditional farmers have often developed strategies for resource management that are not exploitative but serve to enable sustained use over time and may be very efficient with respect to inputs. Hence there are possible models for examination, analysis, understanding and adaptation for modern use.

However, proponents of this approach should not get too carried away in according too much perfection or value to traditional methods simply because they use few inputs other than land and labour or because there is an assumed harmony with the environment. Traditional methods can be exploitative, destructive and anything but sustainable. Destruction of forest and use of fire have been constant factors in the changing status of the vegetation of Papua New Guinea (Bulmer 1982; Powell 1982). Regressive changes are accentuated by increasing population pressure, increases in cash cropping or in commercialisation generally, and by the increased mobility of people consequent to decreasing conflict. These pressures are only partly offset by urban migration and the current political call is "back to the land". If the consequences of a land shortage for subsistence gardens, increased rate of forest destruction, more rapid spread of grasslands, reduced fallow lengths, reduced soil fertility and increased erosion are to be avoided, then there must be increased intensification of production. The easy way out is to intensify through the use of input intensive technologies but such are generally regarded as short-term, unsustainable solutions. The real challenge is to examine and understand the traditional technologies so that the best ideas can be incorporated into a new knowledge about how to achieve *sustainable intensification*. By this is meant the intensive development, protection or use of the renewable resources to enable the maximum number of people and their communities to provide for their social, economic and cultural well-being from generation to generation.

As well as the diversity of food crops, cultural practices and farming systems already discussed, a second feature of agricultural production or development in Papua New Guinea is the dichotomy between improvements that can be made by changes to existing systems and those brought about by replacing traditional systems with new, usually totally commercial, production systems or adding these into the local economy. It is relatively easy to package and extend acceptable technologies for new commercial developments where appropriate inputs and markets are available. Examples are the phenomenal growth of small-scale broiler chicken production with purchased feeds and live bird sales and of marketable vegetable production of mainly introduced species.

It is much more difficult to make deliberate changes to existing systems that will improve output or food security without causing serious disruptions, whether social or biological. Traditional garden production systems cannot easily respond to efforts to increase the marketable surplus and there has been much heart-ache caused to well-meaning extension agents who have attempted to promote this. It is much easier to alienate land for cattle projects or cash cropping, thus causing social disruption and putting more pressure onto the subsistence systems (see Grossman 1984, for a full discussion on these problems with respect to an Eastern Highlands system). Harmonious development can only proceed from a thorough understanding or knowledge of the existing system, how it works and in what respects it might be amenable to productive change.

Sillitoe (1995) gives a recent example of detailed research into a site specific traditional system in his study of the causal agents of organic matter breakdown in local garden cycles and of local knowledge concerning the causes of plant disease and the pests of crop plants. He draws attention to the desirability of incorporating such knowledge into scientific investigations and states: "Further research into soil fauna populations and crop yields is recommended, for it may throw some light onto those constraints that are limiting staple crop yields, and suggest possible interventions to increase them."

Implications for Training

Agricultural diversity requires that much research and development be site specific and it is clear that there is a serious lack of observer knowledge and understanding of the often highly complex traditional gardening systems. For beneficial changes or additions to be made to such systems, new or altered practices must be extended as real solutions to known problems, must meet the expressed needs of the farmers and must not overtax the available resources, including labour. It is necessary to try to understand why previously recommended innovations have not been adopted. Is it that tradition is too strong or is it simply that the suggestions did not "fit"?

Farmers no longer wish to be told what to produce, with projects or schemes designed centrally and extension officers trained to deliver pre-determined messages, organise marketing or monitor input delivery. Indeed, the very concept of "extension" is becoming outmoded since advisory services should be moving away from the concept of simple technology transfer implied by the term. Farmers and landowner or landuser groups need to make their own decisions concerning the best use of the mix of resources available to them, based on the best available information. Hence there is a requirement for multi-way interactions and a participatory approach to resource evaluation, land utilisation and development planning, as well as local on-site demonstration of suggested technological changes. Farmers need assistance from well informed, responsive professionals who can understand the needs and context, source the relevant technical information, interpret it and train the farmers in decision making processes.

If rural development is to be effective, the training of these professionals must include an understanding and appreciation of traditional knowledge, existing systems and social or cultural obligations and constraints. This requires researched information on the problems, needs and motivation of farmers which can also lead on to further research to solve the problems raised and package the solutions for local demonstration.

Agricultural graduates of tertiary institutions may be required to undertake such research or may be filling the communication gap between farmers and research agencies. They will therefore need to be trained for one or other or both of these roles. This means education and training in an understanding and appreciation of traditional practices (including shifting cultivation) and the range of feasible options available to increase production in a sustainable and efficient manner.

There is a perception that traditional knowledge, agricultural diversity and the needs of farmers with continuing involvement in traditional agriculture are being neglected in current agricultural education and training.

While the current situation, at least in the Vudal Diploma programme, is not as gloomy as that assessed by Hill et al (1982) for the Department of Agriculture and Livestock Certificate programmes, the content on these topics in the relevant subjects is still largely descriptive rather than analytical.

The Diploma in Tropical Agriculture syllabi include discussions on the reasons for diversity, the major farming systems, cropping patterns, agronomic practices, the roles of livestock, potentials for change and the likely impact of development, all directly related to traditional agriculture. In addition, the practical work and case studies include sustainable agronomic practices such as rotations, mulching, composting, sloping land management, mixed gardening and other integrated pest management techniques. This is supported by the commercial vegetable production section of the farm, which practises mixed and alley cropping techniques, and by research projects on fallowing systems, organic gardening and the impact of logging and commercial tree crop development on traditional food crop production. Much more research is needed however if even the more prevalent practices and systems are to be understood.

Clearly it is not possible to "research" or "teach" all possible systems or crop genotype x environment combinations. Of course, there are basic principles that apply across all systems, but students who will later be involved in site-specific research and development must not only be educated in basic processes and principles but also taught to apply knowledge to new situations, to observe and interpret, suggest possible options for change, source information, monitor adoption, analyse successes or failures and either research the real problems that arise in the field or relay the needs to appropriate research agencies. Their knowledge and skills would derive from selected in-depth case studies and involvement in a range of good, accepted farming practices. They would need to be able to use a range of analytical tools but, above all, develop as far as possible a flexibility of thinking to help farmers achieve their goals.

Conclusions

The conclusions can be framed in answer to the questions posed in the introduction to this paper. There is clearly a wealth of information regarding crop diversity and the traditional practices and farming systems. There is also clearly a wealth of traditional knowledge possessed by farmers which is site and culture specific. There is a dearth of knowledge however about the mode of action and effectiveness of these practices and the potential usefulness of the diversity of crops and systems. Such knowledge could be of immense value to scientists, advisory officers and farmers engaged in the search for suitable models of sustainable intensification of agricultural production and their application. Graduates in agriculture from tertiary institutions need to be able to expand, develop or use such knowledge to solve the problems and meet the needs of farmers. It is, therefore, necessary that the appreciation and understanding of traditional agricultural knowledge be important components of curricula or subject syllabi of training institutions and students taught to be able to use such knowledge wisely and well.

REFERENCES

Bourke, R.M. (1996) "Papua New Guinea: Text Summaries". Agricultural systems of Papua New Guinea Working Paper No. 1. Canberra: Dept. of Human Geography, Research School of Pacific and Asian Studies, Australian National University.

Bourke, R.M., Allen, B.J. , Hide, R.L., Fereday, N., Fritsch, D., Gaupu, B., Grau, R., Hobshawn, P., Levett, M.P., Lyon, S., Mangi, V. and Sem, G. (1997) "Morobe Province: Text Summaries, Maps, Code Lists and Village Identification." Agricultural Systems of Papua New Guinea Working Paper No. 19. Canberra: Dept. of Human Geography, Research School of Pacific and Asian Studies, Australian National University.

Bulmer, R.N.H. (1982) "Traditional Conservation Practices in Papua New Guinea." In Morauta et al (op.cit.).

French, B.R. (1986) "Food Plants of Papua New Guinea: a Compendium." Australia and Pacific Science Foundation.

________ (1986) "Research Planning for Food Crop Production." Papua New Guinea: Assistance in Establishing a National Agricultural Research Institute (Phase II). Unpublished Report. Rome: FAO-UN.

Grossman, L.S. (1984) *Peasants, Subsistence Ecology, and Development in the Highlands of Papua New Guinea*. Princeton: Princeton Univ. Press.

Gurnah, A.M. (1992) "Agro-climatic Zones of Papua New Guinea." *Niugini Agrisaiens*, 1: 17-22.

Hill, L., Pernetta, J. and Rongap, B. (1982) "The Traditional Knowledge Base: Implications and Possibilities for Contemporary Papua New Guinea". In Morauta et al (op.cit.).

Kambuou, R.N. (1996) "The Current Status and Prospects of Plant Genetic Resources in Papua New Guinea". Country Report: FAO International Conference and Programme for Plant Genetic Resources, Leipzig. Rome: FAO-UN.

Kesavan, V. (1983) "Agronomy: Research, Extension, Training". Appendix III. Profile and Planning Study for Subsistence Food Production, Volume 2: Appendices. Port Moresby: UNDP; FAO.

Macewan, J.M. (1978) "Subsistence Agriculture". Agriculture in the Economy: A Series of Review Papers. Port Moresby: Dept.. of Primary Industry.

Morauta, L., Pernetta, J. and Heaney, W. (eds) (1982) "Monograph 16: Traditional Conservation in Papua New Guinea: Implications for Today". Port Moresby: Institute of Applied Social and Economic Research.

Pernetta, J., Morauta, L., and Heaney, W. (1982) "Overview". In Morauta et al (op.cit.).

Powell, J.M. (1976) "Part III: Ethnobotany". In Paijmans, K. (ed), *New Guinea Vegetation*. Canberra: CSIRO.

__________ (1982) "Traditional Management and Conservation of Vegetation in Papua New Guinea". In Morauta et al (op.cit.).

Sillitoe, P. (1995) "Ethnoscientific Observations on Entomology and Mycology in the Southern Highlands of Papua New Guinea." *Science in New Guinea*, 21(1): 3-26.

Spriggs, M. (1982) "Traditional Uses of Fresh Water in Papua New Guinea; Past Neglect and Future Possibilities." In Morauta et al (op.cit.).

Wood, A. and Humphreys, G.S. (1982) "Traditional Soil Conservation in Papua New Guinea." In Morauta et al (op.cit).

11

Domestic Rabbit Introductions in the Managalase Plateau Area of Oro Province, Papua New Guinea

F. Robinson, I.McL Grant and D.A. Askin.

ABSTRACT

Rabbits were introduced in 1996 to an economically deprived area as part of a conservation strategy for the Queen Alexandra Birdwing Butterfly. An encouragement for villagers to set aside forest reserves for the butterfly was that this allowed them to be involved in rabbit farming. This strategy appears to be successful. Introduction of rabbits appears to be having an impact on both conservation attitudes and on nutrition and incomes of people in an economically deprived area. This report covers the first year of the project.

Introduction

Background: Rabbit farming is of steadily increasing importance in many areas of the world. This is largely due to the rabbit's high rate of reproduction; early maturity; rapid growth rate; efficient feed and land space utilisation; limited competition with humans for similar foods and high-quality nutritious meat (Cheeke 1980). In addition, rabbits have minimal zoonotic health hazards, involve minor capital investment, relative to returns and labour inputs (Mamattah 1978), and are relatively simple to manage. Further, the diminishing supply of bush-meat has been a strong impetus to small-scale rabbit farming.

History of rabbits in New Guinea: Rabbits were first introduced to the island of New Guinea, in what is now Irian Jaya, by the Dutch Administration prior to the Second World War.

Further introductions were sponsored by Australian Baptist Missionaries in the mid-1950s (Mr. Norman Draper, personal communication). Their purpose was to improve the nutrition of protein-poor, isolated Highland peoples. A later and secondary purpose was the formation of small local businesses.

Rabbits are now part of the normal subsistence garden system in the Highlands of Irian Jaya, especially in the North Baliem Valley. They have, however, failed to become as useful in the lowlands, perhaps because the introduced breed (New Zealand White) is not suited to the hot and humid conditions of the tropics. Many people have suggested rabbits would be a useful introduction to Papua New Guinea for both subsistence and commercial exploitation. No successful introductions were made during the time of the Australian Administration, although, several attempts were made (R. Ferraris, personal communication). The dreadful consequences in Australia from the introduction of the wild rabbit precluded all but the bravest from making the observation that domestic rabbits are unlike their wild relative and that the wet tropics are not conducive to the survival of rabbits should they do escape. There are no reports of either feral rabbits becoming established in the wild or of environmental damage in 50 years of experience with domestic rabbits in Irian Jaya (Mr. Norman Draper, personal communication).

Meat rabbits were introduced into Papua New Guinea in 1993 with the aim of establishing a meat rabbit industry among small holders. The initial aim of the project was to demonstrate that under village conditions domestic rabbits could provide meat and income for families, with minimal or no expensive imported concentrates. From the 15 specific pathogen free rabbits introduced, the number of rabbits present in the country has now increased to around 10,000 located at over 800 sites.

Potential of the Rabbit in Papua New Guinea

The rabbit project aims to promote self-reliance among village communities in Papua New Guinea, especially in areas where incomes are low and the people do not have access to regular supplies of animal protein. Traditional systems of agriculture in both Irian Jaya and Papua New Guinea involve a shifting garden system with little intensive animal production. This system is now under pressure as the population rises. A system where rabbits are integrated in sustainable gardening, and agroforestry needs to be developed, as has happened in some areas of Irian Jaya. The growing of legumes, ostensibly for rabbit feed and the use of rabbit manure in kitchen gardens would make site stable gardening feasible.

Oro Conservation Project

The Queen Alexandra Birdwing Butterfly (QAAB) is the largest butterfly in the world. It has a limited distribution, found only in three regions of the Oro Province of Papua New Guinea, and is quite rare. The Oro Conservation Project's (OCP) prime objective is to encourage the Managalase and Plains people of Oro Province to conserve part of the prime habitat of the QABB so that this great butterfly might continue to survive.

The domestic rabbit has been used in trials throughout Papua New Guinea since 1993 under the aegis of the University of Technology Department of Agriculture, and with major inputs from the New Zealand Baptist Mission in the form of staff and expertise. The Managalase project has been part of the overall evaluation. This programme, in association with gardening, health and nutrition education, was part of the overall project strategy for the conservation of the QABB. For the QABB Conservation Project to succeed the whole community must be involved and see that its own interests are served by adoption of an animal, a plant, or an attitude.

Aim

The aim was to improve human nutrition and incomes through improvements in vegetable production in kitchen gardens, and directly through better meat supplies.

Methods

Stages in the introduction of the rabbit in the Managalase plateau were as follows:

Firstly, staff undertook a study of the existing cultural and subsistence agricultural systems. Secondly, two senior staff from the OCP attended a training course in rabbit production and sustainable gardening at the University of Technology in late 1995. Since the prospects for an integrated rabbit and gardening project appeared positive, a central breeding unit was established as part of an awareness programme. This was to show people rabbits and to discuss concerns raised by local village and government people. This activity was general at first so that the whole Plateau had seen rabbits or had at least heard of their presence.

Later, three local villages were targeted as they were easily contacted on foot within one hour of the Project's base station. Families rather than individuals were sought, and there was a conscious effort to recognize that women were the key people in the introduction process. The educational level of the participant was also, but not exclusively taken into account. The selection for the first village trials was made. Ten women from the proposed villages attended the second Rabbit Management and Sustainable Garden Course at the University of Technology in July 1996.

Following their return cages were built and the first six village rabbitries, located in three villages were started. These were visited regularly by OCP staff. Evaluation was made of their viability, feeding regimes, and cultural "fit". The OCP rabbitry at Afore was used as a breeding unit to prepare for further expansion. The central breeding unit provided initial experience for staff of the project learning the daily routine from first hand experience before the possibility of setting up projects elsewhere was canvassed. At this stage had there been cultural, legal, or animal husbandry problems then the experiment could have been fairly painlessly terminated.

Two or three village meetings were held at each of the three villages who wished to participate in the pilot projects. A step-by-step programme was devised in cooperation with the interested villages. The key to this programme was, and remains, a commitment by the prospective rabbit growers, a willingness to undertake training and self-selection. The level of commitment could be judged by the willingness of people to go through the process of "apprenticing" themselves for a short (two week) period. A prospective grower was required to learn basic skills by doing the everyday handling, cleaning and feeding work. This was thought to be a key to the long-term success in this project. There was close and informal contact with the central breeding site. Those students who come through this process were eligible to attend a formal course to learn in detail how to manage their own unit within an integrated garden system. The attendance and completion of the Rabbit Management Course was the criterion for the approval of a project in the village. A Certificate of Completion of this course was given and was, *de facto*, the "license" for obtaining rabbits from the central breeding unit or elsewhere. The final commitment was for the grower to sign an agreement to undertake certain courses of action including the repayment of a loan for the cages and rabbits. Growers could repay the loan either in cash and/or by a tithe system at an agreed rate.

For the second wave of introductions, which involved many more villages, the priority was to ensure that villages were close to a formed road as it was important that there be good access for extension in the early stages of a project. Another criterion was the project's proximity to areas being studied by the OCP Principal Scientist in his work with the Queen Alexandra's Birdwing Butterfly. There was a tendency for the projects to be more numerous in the areas where the butterfly and the parastolochia vine, which provides feed for the caterpillar, are found. The supervision of, and then extension with, the growers are carried out in stages. In the first three months of a projects' existence visits are weekly and up to half an hour in duration. Aspects of management, in particular feeding, were reinforced and further confidence-building skills transfer took place. After the first 3 months visits were made less frequently and depended on the results of previous visits. That is, if farmers were experiencing problems then follow visits were arranged for subsequent weeks.

Results

The response from the local community and individual farmers to introduction of the rabbit was very positive. This resulted in expansion of the programme to secondary sites across the Managalase plateau. Rabbits bred well and there were few disease problems encountered. Farmers fed their rabbits locally available forages, root crops and salt with supplements of copra meal and rabbit pellets. Choice of feed depended on availability.

There were 29 projects established in 14 villages. Twenty-six of these were stocked with 2 does and 1 buck; 3 were stocked with 3 does and 1 buck. From 15 villages with 29 projects (42 does) a total of 78 litters were produced over the period of the study. The average number of kits weaned from each litter was 4.3. The average interval between litters was 13 weeks (range 6-26 weeks). Four litters died because they were scattered on the cage floor at birth instead of being delivered in a nest box. Deaths also occurred due to predation by dogs (5) and snakes (2) and theft accounted for 5.

A bushfire completely destroyed one of the best projects and all kits and breeders on 13 September 1997.

Discussion

In this region, the nutrition of children is generally poor. Many children are 60-80 per cent of normal weight. The introduction of rabbits (in association with a gardening and nutrition programme) might diminish this problem. Hunting of game traditionally supplied meat for families. Hunting now yields far less game in most areas as the population has increased dramatically. The population in 1973 was 6990; in 1990 it was nearly 13,000. Government services have been the normal way for these problems to be tackled, but Government services, including both Agriculture and Health are everywhere in decline. Except in national emergency the reach of the Provincial Government services tends to go no further than the coastal region.

To alter nutrition significantly involves changes to the existing subsistence agriculture system. Changes which are permanent come from within the system rather than being imposed from outside. The rabbit and particularly its manure, if properly integrated within the system, will alter fundamentally some key aspects of the subsistence system allowing for intensification and greater food reliability and variety. This is attainable only if there is a significant adoption over a wide geographical area within the Plateau. The initial aim was to establish at least 50 village projects. This should provide sufficient surplus breeding stock, and local expertise, for all those farmers wishing to establish projects.

Evaluation

From the results obtained by farmers in this region, a farmer with two does should be able to produce about 30 rabbits for eating and sale per annum. These figures are quite acceptable given the low costs of production. This reflects well on the training programme, quality of cages and the level of motivation by the people. There is a tradition of keeping and feeding caged animals and birds preceding European contact and still being practised which may also have contributed to the project's success.

There was, however, considerable variation among farmers in the results obtained. It was observed that some farmers were naturally better at managing their projects than others. In some cases social problems appeared to contribute to poor performance of a project. Lack of confidence in handling stock also appeared to contribute to slow reproductive performance, with farmers reluctant to handle nervous does for mating. With encouragement these projects may improve over time. It appears reasonable to expect that only one-third of village farmers will be interested to make the commitment to farm rabbits and to establish kitchen gardens close to their homes.

Conservation

The objects of a better-balanced nutrition, a better life for the community, and an improvement in the habitat protection for QABB are not mutually exclusive. There is no discernible adverse environmental impact from these activities evident so far. There are however, some indications of a positive impact. The long-term aim is to gain environmentally positive results from the integration of rabbits into the subsistence system. As elsewhere the Managalase communities are based on the family. The work of the OCP focuses on families in the target areas. The family approach means that conservation and the attitudes that go with it are integrated into the normal life of the village. The indications on the Managalase Plateau are hopeful for both nutrition and for conservation of the rain forest.

While it is not true to say that every rabbit eaten is equal to a local bandicoot preserved, just such connections have been made by some Managalase individuals themselves. It is likely that some of the hunting pressure would be reduced if there were more rabbit projects. At this time the scale is not big. In parts of Irian Jaya's Highlands there are hundreds of such projects and a greater positive result is clear. Many workers have observed that if the intensification of subsistence gardens can be achieved then the pressure by an increasing human population on the environment can be reduced. This might, perhaps, be painfully obvious. It is very difficult to achieve this aim in practice. The integrated programme carried out by the Social Development Officer and her team with nutrition, micro-credit for seeds, and the use of appropriate technology (a Chinese rotary-hoe) to promote kitchen gardens, aims to influence people holistically rather than sector by sector. In particular the relationship of the use of rabbit manure on 'Kitchen Gardens' is being explored. This use is a feature of the Irian Jaya projects. Kitchen gardens are used by the Papua New Guinea Highlanders but are still relatively rare in the Managalase Plateau.

Economic Considerations

The economics of rabbits as a small business is attractive to local farmers. They are easier to maintain than intensive chicken production, which relies on high levels of imported feed and stock need to be replaced at the end of each batch. The financial considerations have been discussed fully with the individual project owners prior to their agreement to begin a new project. They are acceptable to the current participants in the programme. Repayments in both cash and by tithes of the rabbits have started. Usually the option to repay half of a surviving litter is taken by the owners. Several individuals have opted to repay in cash.

Additional benefits from farming rabbits are also likely with transfer of knowledge learned to farming of other livestock species. Farmers are interested to develop semi-intensive duck and layer chicken projects under their rabbit cages. This allows better utilization of spilt feed. Insects and worms, which are found in the compost, are also a valuable feed source for chickens and ducks. There is also the possibility of more intensive horticulture through the use of manure and compost. The rabbit owners must deal with both these by-products. The easiest and the most useful way is to incorporate them into a kitchen garden. The use of these materials prolongs the time that the gardens can be used productively and enables high value crops such as cabbage and tomatoes to be grown.

Sustainability

This will depend upon the numbers of viable projects which will be in existence at the time the overseas component of the OCP complete their service and on the ability of the local people to take over organisation of the project. Further growth of the project depends on a constant supply of foodstuffs and cage materials for farmers at reasonable cost. Road access is difficult during the wet season and distribution of materials within the plateau requires some organisation. Rabbits are small animals easy to handle and transport in an area not well served by transport and roads. The cooperation of the Popondetta hardware suppliers in stocking the most expensive of the materials (such as the 0.9cm hard mesh at K12-14.00/m) is possible and likely. The tubing and drinkers are currently purchased from New Zealand Baptist Mission using their economies of scale. Eventually all the requirements will be obtainable commercially.

The sale of rabbits when they become a surplus to subsistence is unlikely to be a problem. Two Popondetta businesses have shown an interest in buying killed and dressed rabbits. One, the Oro Guest House, has put in a regular order when stocks become available, and provided that the sale prices remain comparable with chicken meat.

Important to achieving long-term survival of the project is the establishment of an organisation to foster rabbit production in the region. Fifty-four people met in September 1997 to set up a Growers Group. This will seek further funding as seed money in order to develop the organization and perform some of the tasks currently being carried out by the Rural Development Specialist of the OCP at Afore. Establishment of projects in other regions of Oro province is also important for continuing development. The Popondetta Plains people know of the successful introductions and are keen to begin their own project. There is discussion with them concerning the most suitable methods to satisfy their aspirations. At present there has been only limited assistance with promotion of rabbit farming by the Department of Primary Industry in Oro Province. Some funding has recently been promised to assist with running training programmes for farmers close to Popondetta. Government assistance will be needed for the wider distribution of rabbits in the province.

Gender Issues

Rabbit farming may lead to some enhancement of women's status and more control by them of the food supply. Women mostly carry out the feeding of rabbits. The decisions of breeding, weaning, and the disposal of the compost and the manure will also likely be within the women's role. Selling surplus vegetables at the market is women's work. It is probable that the sale of surplus rabbits will be women's work. The amount of cash per animal is small, but this constant small income will assist women in their family role while offering no threat to the perceived status of men.

REFERENCES

Cheeke, P.R. (1980) "The potential role of the rabbit in meeting world food needs." *J. Appl. Rabbit Res.*, 3(3): 3-5.

Mamattah, N. (1978) "Sociological aspects of introducing rabbits into farm practices." Workshop on Rabbit Husbandry in Africa, Morogoro, Tanzania, 16-21 Dec., 93-98. Stockholm, IFS.

12

Papua New Guinea's Rainforests: Policy, Practice, Stakeholders and Resource Management

Hartmut Holzknecht

ABSTRACT

This chapter discusses forest policies and stakeholder groups in Papua New Guinea. The focus of the discussion is on commercial forestry, which is especially appropriate because it is in this sector that one finds the major negative impacts of such development on leadership, manipulation of the majority of resource owners and short-term financial benefits for a few to the detriment of long-term resource sustainability for the majority. The paper emphasises the importance of customary resource ownership and argues that there is a current critical need to build on the strengths of these customary systems. Since virtually all resources in Papua New Guinea are owned and controlled through customary processes it is only through such a focus that appropriate and forward-looking community resource management approaches in Papua New Guinea can be encouraged. To ignore such customary systems and to continually try to bypass or undermine them is not only to act against The Constitution, but also to act against the great majority of the population which has indicated a strong preference to maintain such customary systems.

Introduction

The coherent continued existence of Papua New Guinea's environment for basic human survival, for biodiversity conservation and as a sustainable source of livelihoods of various kinds has been much debated both within Papua New Guinea and overseas. A critical aspect of this continued existence is appropriate resource management. This paper[1] presents a brief overview and analysis of aspects of the development of policy and practice in relation to natural resource management in Papua New Guinea. A reform process has been underway in the forestry sector since the late 1980s and a number of positive results have begun to emerge. Although the forestry sector will be the focus of this paper, the conclusions reached are applicable across a range of natural resources.

Papua New Guinea as a nation became independent in 1975 and is made up of a multitude of smaller and larger ethnic groups, speaking at least 700 languages. The majority of these groups were egalitarian in structure, that is, non-hierarchical. Leadership was earned by a person's activity and personality against sets of ideal criteria (varying from place to place) for leadership. A wide range of diverse customary land tenure and land use systems (see Crocombe and Hide 1987) continue to exist for allocating rights to resources and for regulating the succession to these rights. In most areas, Papua New Guinea societies were classless and casteless and all the population had access in varying ways and degrees to land and resources for subsistence needs, shelter and the other necessities of life. These systems ensure that rights to all resources are privately owned, some rights by individuals and other rights by corporate groups (such as clans, sub-clans and lineages).

Some 3 per cent of the land area is now alienated to the State. Customary resource rights, legal systems and 'Papua New Guinea ways' are recognised by and enshrined in the National Goals and Directive Principles set out in the preamble to *The Constitution* (Papua New Guinea 1975). Past colonial and more recent State attempts to regularise land tenure and land ownership in Papua New Guinea have floundered due to a lack of clear understanding of the far-reaching implications of these recognised customary systems. The result is that these customary resource systems are, by and large, still in place in spite of determined efforts to break them down or bypass them. The modern formal and constitutional recognition of such rights and customary resource tenure and use systems is, other than in other Melanesian countries in the South Pacific, virtually without precedent.

A brief clarification of resource rights is needed here: a wide range of rights to land and other resources is, in most cases, vested in customary social groups such as clans. Individuals may exercise all or any of these rights, either by inherited membership of such a kinship group (through patrilineal or matrilineal or other means), or by accessing temporary use rights through various kinship or friendship links. Those inheriting such rights from their ancestors may be termed group members with 'permanent' rights while others can only access 'temporary' rights of use of land or other resources. For example, temporary gardening rights may be in effect only for the life of that garden.

In pre-colonial days and in the colonial period there was some considerable degree of flexibility possible for outsiders to become incorporated into corporate groups, such a move requiring residence in, and commitment to, the new group (see Watson 1980). Societal mechanisms existed (and mostly still exist) for incorporating refugees from warfare, orphans and other adoptees into existing social groups. Initially such newcomers were treated as quasi-kin and then, over time and through other mechanisms (such as intermarriage), they could become classified as true kin. Particular kinds of rights are always held by individuals — this applies especially to the ownership of economic trees (betel nut palms, breadfruit, coconuts, *Canarium* and *Terminalia* spp., *Pandanus* spp., etc.).

The Context of Policy Reform

In response to widespread public concern about the situation in the forestry sector, the Papua New Guinea Government in May 1987 set up a Commission of Inquiry into the Operations of the Timber Industry under Judge T. Barnett. The main findings of the Barnett Inquiry are set out below:

Main Findings of the Barnett Inquiry[2]

- widespread corruption
- regular breaching of permit conditions
- comprehensive transfer pricing practices;
- regular misdeclaration (therefore undervaluing) of species
- inadequate or non-existent monitoring
- environmental destruction through bad logging practices
- unsustainable logging activities
- manipulation and cheating of resource owners

The Barnett Inquiry began the push for reform in forest policy and practice. Much of this has taken place through a National Forestry and Conservation Action Programme (NFCAP) with the commitment of various donors to support a range of research and development projects within NFCAP. A new National Forest Policy, developed within the Department of Forests, was unveiled in 1991. The National Parliament approved a new *Forestry Act* in 1991 (not gazetted until 1992 with implementation beginning only in 1993), repealing amongst others the previous *Forestry Act* and the *Forestry (Private Dealings) Act*. The effect of the amendments was to drastically reduce the ability of the government to control and bring timber companies into line in the 75 then active logging projects around the country by the last-minute insertion of a clause exempting all current logging permit holders (almost all of them foreign logging companies) from complying with all new 1991 *Forestry Act* requirements. The National Forest Service, within a Papua New Guinea Forest Authority, began to be established in mid-1993; 75% of staff were to be project and field-based officers.

Forest Development Guidelines (see Holzknecht 1994, 1995 for more details) were prepared and endorsed by the National Executive Council in December 1993. Many of these proposals and activities met with active and increasing opposition, especially from commercial interests in the foreign-dominated industry. During the latter half of 1993 and early 1994 in particular there was a concerted campaign in the media against a number of the provisions set out both in the new 1991 *Forestry Act* and the subsequent Forest Development Guidelines. The result was a suspension by the State of the implementation of all forest development guideline proposals for a time. The proposed new revenue system was scrapped and a new revenue system was put in place through the 1996 budget with increased revenues, especially earmarked for resource owners, beginning in 1997.

Stakeholders in the Forest Sector

There are a number of stakeholders in the forest sector, each with their own agendas and ways of working; they include resource owners, landowner companies, the State, logging companies and/or contractors and non-government organisations. I discuss them in turn and set out a number of relevant aspects, including their incentives to conserve, degrade or properly manage the forest.

Resource Owners

Outsiders tend to see this category of stakeholder as an undifferentiated group and so the ubiquitous term 'landowner' has achieved almost mystical or 'noble savage' proportions, to be challenged at the challenger's peril. In fact this category consists of a number of subcategories.

(i) Permanent Resource Rights Holders

Some aspects of this and the next subcategory were discussed above. This subcategory, like the next, is not necessarily a unified group. Anecdotal evidence from many locations where logging projects are operating suggests that old people in particular pushed for logging to take place so that they could enjoy some of the benefits of royalty monies before they died. Women, in particular, have less control and influence over decisions made concerning resources. However, they are the part of the community which makes more regular and consistent use of the forest as a forage, fuelwood and resource base for a wide range of activities; they and the young are the ones who suffer first in trying to maintain a reasonable living standard for the family in the face of the dramatic changes brought about by commercial logging to the resource base.

(ii) Temporary Resource Rights Holders

The category of temporary resource rights holders can also cover a very wide range of persons and situations which vary widely from area to area and therefore are impossible to enumerate in detail here. Suffice it to say that population pressures on one or more natural resources play a significant part in how wide the network of individuals is (whether kin or not) which feels it has some kind of temporary access or has made arrangements to have access to resource rights. In the Highlands of Papua New Guinea, for example, where population density tends to be much higher than in the lowland and coastal areas, the range of people gaining temporary access to resources can become more and more limited than in the other areas, as the population increases.

Leading Players

These individuals may belong to one or neither of the above two subcategories, but are likely to be in very influential positions, such as elected members of a provincial government, or elected members of the National Parliament (see 'landowner companies', below), or senior public servants. Because their positions require them to spend a great deal of time in urban areas, especially in Port Moresby, these leading players are on the lookout for resource development opportunities. Initially, at least, intentions in most cases are honourable as they try to find ways of assisting their home areas to achieve 'development'. They are, in turn, vulnerable to flattery, self-interest and 'capture' (the myriad of ways are well-known and need not be detailed here); once they are in such a patron-client relationship,[3] the process of "enclosing the commons" (Hardin 1968, Ostrom 1990, Holzknecht nd) for individual benefits begins. The leading player's view then, supported by a small clique of close relatives and other hangers-on, tends to be of short-term pecuniary and power self-interest, with community rights (whether permanent or temporary) getting short shrift or being at least temporarily placated by promises of future wealth and assistance.

Landowner Companies

So-called 'landowner companies' are characterised by their shareholders and management tending to be mostly of one faction or one part of a group, all supporters of a particular leading player such as a provincial politician or a national Member of Parliament who has made contact with, or been contacted by, an overseas logging company. A 'landowner company' is an exploitation strategy which began to be developed some ten years ago, originally intended to represent all customary resource owners in a particular area.

Once the contact and the promises have been made (behind closed doors) a company is formed and registered with this core group in control. The landowner company then agitates at the highest circles for it to be granted the logging permit. When this has been successfully achieved the original foreign logger is then contracted to log and market the timber from that concession. Other than the direct royalty payment which is made directly to a list of individuals, the remaining funds are captured by the landowner company to do with as its principals please. Most landowner companies rarely hold public meetings, much less the statutory annual general meetings at which office-bearers are normally elected. Very few send in their statutory annual returns to the Registrar of Companies, very few present financial reports of income and expenditure to the mass of resource owners in the concession area. Management officers are usually not active at home and usually reside in the provincial or national capital, often leading extravagant life styles.

The *Forestry Act* (1991) requires such landowner organisations to be fully representative of the different groupings of resource owners within a concession area. Such resource owner organisations are required to apply to the Papua New Guinea Forest Authority for recognition and registration by tabling a number of documents to prove their representativeness. A number of the more vocal "landowner companies" continue to object to these requirements, but this can be best described as a rearguard action by those entrenched in power and in control of logging revenues, carried out to protect themselves.

The State

The State has a direct interest in the forest sector from two points of view. Firstly, by act of Parliament, the State controls which non-customary person or private enterprise has access to the timber resource (if the customary resource owners agree and if it is to be harvested commercially) and in what manner and through the approval procedures. Through this licensing system the State directs timber permit holders regarding volumes of timber to be harvested, mitigation of impacts, minimum diameters of logs to be cut, minimum standards of infrastructure and other requirements to be put into an area and reporting procedures. The State controls the level of royalty payable to recognised resource owners. This first element of control is carried out through the Papua New Guinea Forest Authority, a statutory body.

Secondly, the State collects a number of different fees from such licensed operators, including log export taxes. This element, a significant proportion of the State's annual revenue, is basically controlled and collected by the Department of Finance and does not accrue to the Papua New Guinea Forest Authority itself. The State interest in income derived from the forest sector should in theory be balanced by a major State interest in measures to ensure not only that its requirements are being met by industry operators but also that the resources themselves are carefully harvested and are able to properly regenerate themselves so that at some future date they may again be harvested. Until quite recently the monitoring element for these activities has been almost entirely missing, allowing a virtual 'free-for-all' approach by logging contractors, as was highlighted by the Barnett Inquiry.

In addition, any practical interest in regeneration of natural forests and in the establishment of plantations, despite some impressive research results in different parts of the country (see Cameron and Vigus 1993, Vigus nd), has in practice been perfunctory at best. Another balancing aspect to State income derived from various taxes and levies on the forestry sector should be broad-based benefits for the population at large, such as improving infrastructure and services (Holzknecht 1997b) though this, too, leaves much to be desired.

Logging Companies and Contractors

As Judge Barnett found (speaking here particularly about New Ireland Province in the mid to late 1980s),

> It would be true to say of some of the companies that they are now roaming the countryside with the assurance of robber barons; bribing politicians and leaders, creating social disharmony and ignoring laws in order to gain access to rip out and export

> the last remnant of the province's valuable timber (Papua New Guinea 1989).

In reality the situation has considerably worsened since that time. While policy developments and the reform process give hope that there will be rather more overall control, more detailed monitoring of activities as compared to new standards of logging activities (see Papua New Guinea FA/DEC 1995), and more attention paid to scrutinising logging proposals (including detailed working plans and environmental and social impact aspects), the fact of allowing exemptions from the new legal requirements to all existing logging permit holders makes a mockery of much of the 1991 *Forestry Act*.

Logging companies have, both in the past and at present, found sufficient ways of bypassing various problems. By linking up with a local leader and entrepreneur and a group of his supporters and ensuring that this individual and group are, in various ways, kept happy, loggers have created a buffer between themselves and the majority of resource owners in an area. When there is a patron-client relationship[4] (already mentioned above) the effect of this is to deprive most resource owners of a continuing involvement in their forest resource. The patron-client relationship diverts expected and promised profits through a "landowner company": and thus into the hands and control of a small minority of individuals and it also removes from the majority of resource owners any say in the use of such profits. Whenever there are so-called "problems with landowners," legitimate questions being asked and the logging company being challenged, the "landowner company" executives are the buffer and take the side of the logging company. The usual response is to call in the police[5] to harass the protesting resource owners.

The interest of logging companies is to harvest as much marketable timber as possible from a concession at the least cost and with the least trouble. Though royalties to resource owners and levies of various kinds to the State may be paid at a particular time, logs are often stockpiled for a longer period to await a higher world price. Most logging companies spend as little as possible on their own infrastructure in a logging concession, bring in most of their personnel (trained or otherwise and often in contravention of Department of Labour requirements) and second-hand equipment from overseas. Relatively few Papua New Guinea nationals are employed by logging companies and then mostly as unskilled labour.

Many logging companies which have operated in Papua New Guinea diversify their activities to a certain extent by setting up trade stores or transport operations. One particular group of companies, besides controlling some 80 per cent of log exports from Papua New Guinea has diversified its activities into real estate, trading and media.

Non-government Organisations

Non-government organisations (NGOs) have begun to play an increasingly influential role in the forestry debate in Papua New Guinea. The NGO movement in the country is relatively young and becoming more active in areas which have been either abandoned (such as literacy training) or very poorly funded by government.

Like NGO's everywhere, they are responding to increasing international trends which demand community participation in decision-making processes, demand that environmental concerns are taken account of and that integral human development be the first priority in any kind of development. Papua New Guinea NGOs can to a large extent be influenced by international NGOs in terms of stance[6] and tactics and are to a greater or lesser extent dependent on them and other external sources for funding.

As government services and activities have begun to be withdrawn more and more from remote and outlying areas so rural individuals and communities have begun to turn to NGOs such as environmental NGOs, church organisations and others for advice, assistance and information. One NGO, the Individual and Community Rights Advocacy Forum (ICRAF), for example, is playing a vital role in advising resource owners of their legal rights and how to exercise those rights. ICRAF also provides legal aid and backup for cases which resource owners wish to take to court (ICRAF nd).

With a seat on the Board of the Papua New Guinea Forest Authority, the Papua New Guinea NGO movement is in a position of some influence at this level. With regard to resource owners (who tend to be based in rural areas) NGO's have been sporadically carrying out campaigns focussing on such areas as public awareness and literacy. As always, funding these, as well as just keeping these organisations going, is a major battle. They do their best to bring about an awareness of the importance not only of forests but also of all aspects of the environment in the lives of subsistence farmers and to encourage people to think about alternative ways of finding cash income. In this they tend to be let down by their own lack of practical information about alternatives and options and related implications. To raise awareness without being able to encourage people to consider viable options leads only to their frustration and anger.

Papua New Guinea NGOs have also begun to use their international constituency and regularly appeal there for support and in order to extend their own public campaigns internationally. This can be an advantage but can also be a problem in that international NGOs often see such campaigns in simplistic terms and regularly get their information not quite right. Papua New Guinea NGOs, besides (or because of) having continual funding problems, also suffer from not having sufficient numbers of trained and experienced personnel.

Discussion and Alternative Approaches

This paper has discussed the context of customary tenure, policy reform and provided some detail on the stakeholders in Papua New Guinea's forest sector. This discussion makes it clear that it is risky to approach this sector naively and simplistically, for simplistic it is not.

Two themes come through the data which need to play a more positive role when searching for alternative approaches. These are, firstly, the continuing importance of customary resource tenure and resource use systems in Papua New Guinea and, secondly, the need to return resource management strategies and the related responsibilities back to resource owner-based groups and communities across Papua New Guinea. I discuss these in turn.

Continuing Importance of Customary Resource Systems

As the marches, riots and other disturbances in Papua New Guinea in the second half of 1995 showed,[7] Papua New Guinea citizens are not willing to allow their customary resource tenure and use systems to be controlled and rigidified by any proposed legislation. These customary systems are recognised by *The Constitution* and have survived previous colonial and post-colonial attempts to bring them to book and law. These proposals would have frozen these systems in time and so removed one of their essential and necessary characteristics, namely, their flexibility, and so their ability to respond to changing situations and circumstances.

Nevertheless, some developments acceptable to rural resource owners throughout Papua New Guinea must soon take place within these systems,[8] developments which build upon the strengths of these customary systems and use these strengths to continue to maintain and enhance the integrity, as well as the responsiveness, of these systems. For its time a very progressive and innovative piece of legislation, the *Land Groups Incorporation Act*, was passed by the House of Assembly in 1974 following recommendations from the Commission of Inquiry into Land Matters mentioned at the beginning of this paper (Papua New Guinea 1973). This Act built on Papua New Guinea customary corporate groups (such as clans) which controlled and still control access to customary land and resources.

The Act assists in identifying and establishing such customary social groups which exercise such controls within a particular area. The Act gives this unit recognition and standing under modern Papua New Guinea law (subject to certain conditions and ways of incorporating, registering and to certain ways of operating). What emerges out of this process are 'incorporated land groups' (ILGs) which, for different purposes and in different domains and activities, may also still continue to be the customary social group (such as a clan or subclan).

The Act and Incorporated Land Groups

A land group is "that group of people empowered by custom to make fundamental decisions about their land without reference to any other group" (Power 1996:273). Land groups incorporated under the *Land Groups Incorporation Act* are customary land and resource owning groups which operate under custom and are recognised by law. The Act requires a land group to formally identify itself by listing current members and membership criteria (these being supported by genealogies). Membership criteria need to be agreed upon, including that members need to identify themselves as such but also be recognised by other members as legitimate members. Membership lists can be reviewed and revised on an annual basis, thus maintaining the essential flexible character of customary groups. The land group is required to formally identify its properties, all named by custom and known intimately to clan members and neighbouring clans and verified by consensus with these neighbouring groups.

The land group is required to have a constitution setting out a number of matters. This can be in as much or as little detail as the group wishes keeping in mind that custom will apply if a matter is not set out in a constitution. The land group's committee may not make major decisions without the approval of an assembly of group members. In relation to very critical matters, more than one such assembly is required. All major decisions must be recorded in writing. In addition, land group committees may not action major initiatives by themselves but only by active consent of the group's members. Regular meetings and consultations between committee and members will require any dealings to remain transparent, and information on incomes and expenditures to be accessible and shared. In relation to land itself, the Act does not of itself register land in the land group's name in the accepted sense. However, it does record the land group's interest in the properties which are listed in its constitution. Land groups will usually entrench a legal requirement in their constitutions which disallows any kind of sale or alienation of the land group's properties (except under very specific and special circumstances).

The *Land Groups Incorporation Act* of 1974, therefore, approaches the issue of control and management of land and resources in an innovative and constructive fashion without losing sight of the major aim of maintaining the customary strengths of resource tenure and use systems. For example, not one of the above requirements interferes in any way with the temporary access rights to land and resources which are arranged by custom for various purposes and periods with permanent land group rights holders by individuals who do not have such customary rights of access and use.

A criticism often heard from the mostly foreign-dominated business sector in Papua New Guinea is that the continuation of customary resource tenure and use systems acts as a heavy dampener on entrepreneurial activities and on investment and can provide no security. In contrast to this criticism, the recognition of ILGs will, with all modern legal securities, allow them to make arrangements both with their own members and outsiders to use a parcel of land (for example), for a particular purpose, for a specific period of time and under certain conditions (including limits on its use and environmental factors). Lakau (1995), too, brings together data on the productivity of customary resource systems which undermines this criticism, for example, in the proportion of Papua New Guinea's annual coffee production which is grown by smallholders on customary land.

ILGs have already been introduced in a number of mining projects as a way of organising customary resource owners, in the first instance to receive royalties and compensation payments. In addition, the 1991 *Forestry Act* requires all new forestry projects, through an awareness and information campaign to assist resource owners to organise themselves in line with the features of ILGs set out above; this process has now been underway since mid 1995.

Community Based Resource Management

The creation of ILGs cannot be seen as an end in itself, just as a useful mechanism for government and foreign companies to expedite the setting up of resource exploitation projects. However, the process should not stop there. An ILG is, in effect, the first step in a a much longer-term process which recognises that ownership and management control lie with resource owners and resource owner groups. Only they themselves can become more actively engaged in the proper and long-term management of all their resources for the benefit of the whole community and not just for a privileged or powerful few (Holzknecht 1995). The State needs to find appropriate ways in which to facilitate the support of customary systems to help them to become more effective.

One factor which has severely limited resource owners' decision-making is access to information regarding alternatives or options and the implications of such choices with regard to the management, development and conservation of their resources. This area of resource owner access to information, in particular, will need to be solved very soon if a community-based resource management alternative to the current free-for-all in resource exploitation is to gain any credibility and if this alternative is to show sustainable results. Another limiting factor is customary resource owner access to support services such as technical advice, finance and marketing.

Issues related to the long-term promotion of a successful community based resource management process include facilitation training, development of frameworks within which a number of relevant processes can be worked out which are nationally consistent but locally adapted, the levels of cooperation within a community, the needs for local natural and human resource surveys as basic necessary starting positions and the development of appropriate support systems for community based resource management (see Holzknecht 1997a:379-384).

Concluding Remarks

Papua New Guinea's policy reforms are currently beginning to bring some kind of order back to the forestry sector. In addition, the 1991 *Forestry Act* (as amended) has taken its lead from *The Constitution* in requiring the involvement of customary resource owner groups in this process. The State to a large extent still appears to be in thrall to foreign logging interests and dependent on them for the export and marketing of its timber and also for the income it derives from the export of timber.

The approach presented in this paper, that of ILGs, coupled with a community resource management focus is realistic and achievable, given that in a customary and general sense this is already what is taking place, with more or less success, across Papua New Guinea. ILGs are being used and are beginning to have an effect in the mining industry as a productive way to regulate the relationship between mining company and customary resource owner groups. ILGs are also beginning to be applied in forestry projects. Here, as in the mining sector, the majority of problems which have emerged can be traced back to two sources. These are, firstly, government or company staff taking shortcuts and hurrying through the identification and incorporation process for ILGs in a particular area and, secondly, the problems that arise when individuals within a group attempt to shortcut procedures to ensure that only they and their support groups gain whatever monetary benefits come out of the ILG process and the resultant relationship with a development company. This latter is just a continuation of the existing style of patron-client relationship in which one person, or a small clique, try to benefit themselves at the expense of others who have customary rights in a particular area.

This chapter, while it has not been able to do justice to the large and complex topic of customary resource tenure and use systems, has shown that the strengths of these systems, and the wishes of customary resource owners in Papua New Guinea, require that they be taken into account. Thus, customary resource owners in their representative groups and in their communities need take increasing responsibility to ensure that their role in broader based, longer term and sustainable resource management is understood, recognised and supported.

NOTES

1. A somewhat different version of this paper appeared as Holzknecht (1996).
2. More details on this can be found in Barnett (1992), Papua New Guinea (1989) and Holzknecht (1994).
3. The irony of such patron-client relationship in this context is that the client here believes he has the patron as a client and acts accordingly. In reality, the patron well and truly controls the purse-strings and so the relationship and the real financial outcomes of the relationship.
4. See Dauvergne (1995) for details on how such relationships develop in relation to resource exploitation in other countries.
5. Logging companies, in various ways, ensure that there are excellent relationships between themselves and the provincial police command and other provincial authorities.
6. The term 'wildlands', for example, has some meaning in environmental rhetoric, particularly in the US, but is quite meaningless in the Papua New Guinea context. Here people are always part of the landscape and must be taken account of in any environmental planning and conservation measures.
7. These disturbances came about largely as a result of an NGO-led popular movement. This movement initially demonstrated against a component of the World Bank funded Land Mobilisation Project. Without going into the rights and wrongs of these demonstrations, they were based on the belief that the World Bank was pushing Papua New Guinea into individualising land tenure in the country or, alternatively, that the World Bank was intending to take away all customary rights to land and other resources. This movement then continued

demonstrating against a number of the clauses or requirements contained in a World Bank-led Structural Adjustment Programme which was offered to (and accepted by) the Government of Papua New Guinea as a means of rescuing it from a very difficult economic situation.

8. Using Hooper and Ward's phrase (1995:250) we can say that Papua New Guinea currently has a 'breathing space' in terms of its customary land tenure and land use systems. Lakau (1995:15-22) has documented the various attempts made under colonial administrations, as well as more recently, to adapt or change these systems. Such attempts consistently bypassed, ignored, or (at worst) undermined the strengths of the customary resource systems described in this paper and elsewhere (see also Holzknecht, nd).

REFERENCES

Barnett, T. (1992) "Legal and administrative problems of forestry in Papua New Guinea," in *Resources, Development and Politics in the Pacific Islands*, edited by S. Henningham and R. May. Bathurst: Crawford House.

Cameron, A.L. and T. R. Vigus (1993) *Regeneration and Growth of the Moist Tropical Forest in Papua New Guinea and the Implications for Future Harvest*. Brisbane: CSIRO.

Clarke, W., Crocombe, R., Lakau, A., Larmour, P. and McIntyre, M. (1995). *Customary Tenure Systems in Papua New Guinea*. Canberra: AusAID.

Crocombe, R. and Hide, R (1987) "New Guinea: Unity in diversity," in *Land Tenure in the Pacific* (3rd ed), edited by R. Crocombe. Fiji: University of the South Pacific.

Dauvergne, P. (1995) *Shadows in the forest: Japan and the politics of timber in Southeast Asia*. PhD dissertation, University of British Columbia.

Hardin, G. (1968) "The Tragedy of the Commons," *Science* 162:1243-48.

Holzknecht, H. A. (1994) "Managing Papua New Guinea's forest resources," *Development Bulletin* 31: 27-31.

Holzknecht, H. A. (1995) Working Papers Produced as Part of the FMPP Landowner Involvement Component 1993 to 1995. Port Moresby: Papua New Guinea Forest Authority, Groome Poyry Ltd. and World Bank.

Holzknecht, Hartmut (1996) "Policy Reform, Customary Tenure and Stakeholder Clashes in Papua New Guinea's Rainforests," Network Paper, 19c:1-20, Rural Development Forestry Network. London: Overseas Development Institute.

Holzknecht, Hartmut (1997a) "*Pro Bono Publico*? Conservation and Appropriate Resource Management Strategies in Papua New Guinea," in *The Political Economy of Forest Management in Papua New Guinea*, edited by C. Filer. Monograph 32. Port Moresby: Papua New Guinea National Research Institute and London: International Institute of Environment and Development.

Holzknecht, Hartmut (1997b) "Two Sides of the Coin: the case of forestry," in *Compensation for Resource Development in Papua New Guinea*, edited by S. Toft. Law Reform Commission and Pacific Policy Paper 24. Canberra: National Centre for Development Studies, The Australian National University.

Holzknecht, H. A. (n.d.) "Customary Property Rights and Economic Development in Papua New Guinea," To appear in a volume edited by T. van Meijl and F. Benda-Beckmann. London: Kegan Paul.

Holzknecht, Hartmut and Kilyali K. Kalit (1995) "Forest resources: what hope for the future?" *Pacific Economic Bulletin* 10 (1): 95-100.

Hooper, A. and R. G. Ward (1995) "Beyond the Breathing Space," in *Land, Custom and Practice in the South Pacific*, edited by R.G. Ward and E. Kingdon. Melbourne: Cambridge University Press.

ICRAF (nd) *Law for Landowners*. Gerehu: Individual and Community Rights Advocacy Forum Inc.

Lakau, A.A.L.(1995) *Customary Land Tenure, Customary Land Owners and the Proposals for Customary Land Reform in Papua New Guinea*. Lae: Pacific Land Tenure and Policy Centre.

Lakau, A. (1997) "Customary land tenure and common/public rights to minerals in Papua New Guinea," in *The Governance of Common Property in the Pacific Region*, P. Larmour (ed). Pacific Policy Paper No. 19. Canberra: National Centre for Development Studies, and Resource Management in Asia-Pacific Project, Research School of Pacific and Asian Studies, The Australian National University.

Ostrom, E. (1990) *Governing the Commons*. Cambridge, Mass: Cambridge University Press.

Papua New Guinea (1973) *Report of the Commission of Inquiry into Land Matters*. Port Moresby: Government Printer.

Papua New Guinea (1975). *The Constitution of the Independent State of Papua New Guinea*.

Papua New Guinea (1989) *Commission of Inquiry into Aspects of the Forest Industry. Final Report* (2 volumes). Port Moresby: Department of the Prime Minister.

Papua New Guinea Forest Authority/Department of Environment and Conservation (1995) "Papua New Guinea Logging Code of Practice - Final Draft." Boroko: Papua New Guinea Forest Authority and Waigani: Department of Environment and Conservation.

Power, A. P. (1996) "Sustainable Forestry in Papua New Guinea," in *From Rio to Rai: Environment and Development in Papua New Guinea up to 2000 and Beyond*. (Papers from the 20th Waigani Seminar). Volume 3. *A Quarter of Next to Nothing*, edited by D. Gladman, D. Mowbray and J. Duguman. Port Moresby: University of Papua New Guinea Press.

Vigus, Tom (n.d.) *Reforestation Naturally! Kimbe: Kandrian Gloucester Integrated Development Project*. Canberra: Project Design and Management and ANUTECH.

Watson, J. B.(1970) "Society as Organised Flow: the Tairora case." *Southwestern Journal of Anthropology*, 26(2): 107-124.

13

The Option of Surviving Intact — People and their Environment: Researching Shelter in Rural Papua New Guinea Villages

Richard Burnham, Cletus Gonduan, Andrew Sariman

Introduction

This chapter describes issues that concern the well-being of the Papua New Guinea environment and people living in rural areas. It deals with how people and the environment intermingle and how people work with or against the sustainable natural world.

As today's capitalist global economy expands and seeks further markets for resources, consumers and workers (Wallerstein 1984), it is reasonable to suppose that linkages between rural dwellers, city dwellers, and the global economy will not only be established but will strengthen over time. Part of new global linkages and drives for expansion is the shift in many parts of the world from a dependence on traditions to more modern ways. As the global economy continues to expand it is foreseeable that new pressures will be brought to bear on rural people living sustainably with their environment. For some this new intervention — this shift in villager focus — can be tragic, but for others it can result in an astutely managed village that protects its relationship with the sustainable natural world through selective participation in the global economy.

As non-sustainable manufactured goods from the global economy are introduced into the village, the environment's relationship to village life can change from a symbiotic partnership to that of an exploitable commodity. This new perspective of extraction for cash can devastate the sustainable environment and intertwined traditional ways of living. And it can happen most anywhere, especially in the so-called "developing world,"[1] where the desires of cash-poor people for the unaffordable products of the global economy are dammed but not dissipated. How this happens isn't difficult to see in Papua New Guinea. Forests have been clear cut and reefs dynamited and poisoned for cash production.

These acts can undermine the balance that exists between the village and the surrounding environment. When the village pillages the environment to mollify desires only satisfied through cash transactions, the environment might no longer provide subsistence or even sufficient natural or cropped goods for market. It is probable, then, that urban migration could be forced upon some villagers.[2]

But actions that respect the environment and village well-being that knowledgeably integrate humans in the complexity of both the sustainable world and the global economy are possible. In the domain of building materials selections for housing, it is not an *a priori* that non-sustainable materials are anathema for village well-being. Wise decisions by villagers can successfully integrate non-sustainable building materials into the traditional rural village when the complexity of positive and negative factors is unravelled and made clear to village people. The intent is that the environment is to remain vibrantly alive while the people learn the consequences of their desires and act in the best interests of the sustainable village and the environment.

Our research will identify and explore the relationship between the environment, building construction practices, actual building materials performance as compared to villager expectations, cash requirements, and villager time demands associated with various materials. Utilising a case study approach we will develop data in a number of villages that will be representative of rural settlements over a number of broadly defined different environmental niches in Papua New Guinea. When an appropriate number of case studies have been completed, we will develop guidelines for environmental conservation for rural villages which will combine sustainable living practices with an appropriate participation in the global economy. A key aim is to prepare and make widely available an empowering methodology that can be understood and used by villagers in their decision-making processes.

Village Housing

Housing is the overridingly prominent feature of the Papua New Guinean village. Along the southern Huon Gulf coast and in many Milne Bay Province islands, for example, rural villages are made of grouped houses (Figure 13.1) of variable density that, traditionally, are built of local materials. Many houses make use of sago, nipa, pandanus palm fronds and stalks, the coconut palm, a great variety of local woods, bark, and vines. These natural materials are often improved by adding value through such methods as forming sago fronds with presses (Figure 13.2) and the making of roofing panels (Figure 13.3) or impregnating wood with salts from the sea.[4] These materials usually come from the customary owned land of the village. Material may be harvested at a rate that does not impinge on new growth thus providing sustained use over time. The physical outcome is what many take to be as typical — and very often distinctively sensitive and elegant — village architecture (Figures 13.4, 13.5, 13.6, 13.7, 13.8). It is an example of a capable humankind knowledgeably enmeshed in the natural environment.

Yet villagers returning from urban areas may bring home new cultural values that, like new dance steps or new songs, are passed on to others as something superior to the old. Sheet metal roofing, steel nails, plastic sheeting, glass, steel pipe, plywood, milled lumber, and commercial paints are among new materials that replace the traditional materials of the village (Figures 13.9,13.10).

If such values are accepted, then the village takes steps toward different, nontraditional, building materials that make a new sense to villagers; that create a built environment that is removed from the sustainable natural environment. Such a shift might logically continue until sustainable local materials are nearly or completely replaced (Figures 13.11, 13.12). The outcome of this scenario has not only cultural significance (to be addressed below), but it would signal a separation of Papua New Guineans and their dwellings from the natural environment and an abandonment of one traditional pattern of natural resource use (Figure 13.13).

Cash Production by Extraction

Villagers don't buy pandanus or sago for roofs. Nor do they purchase mangrove logs for house pilings. They harvest these and many other sustainable building materials from their land. But for nonsustainable materials they must pay with cash. As a villager not connected to a wage-earner's job, the means to cash is to harvest cultivated crops, or to harvest the natural environment. Cultivated crops such as cocoa, copra, coffee, and betel nut can bring in substantial cash benefits. Garden crops such as taro, banana and peanuts earn less cash. But for those without sufficient crops under cultivation or access to markets for cultivated crops, the option is to harvest from the natural environment. When villagers in need of cash take fish from coastal waters, offer local timber to logging concerns, and take beche-de-mer from the sea floor along their coast, they are utilising the natural environment's products as commodities.

Heavy extraction from nature for cash production often results in disastrous environmental effects and can disrupt village life. The selling of timber rights can result in clear cutting that can lead to erosion and unsettle the ecology of the area. Over-fishing of reefs can result in insufficient protein for village consumption. These and other acts of over-harvesting, designed to produce the required cash that may be used to purchase nonsustainable building materials, can result in sufficient environmental damage that the life of the sustainable village is placed in jeopardy.

Young adults may be forced to migrate from the village to urban environments to ease the pressure on elders still living with a diminished environment, and to earn money for remittances to be sent or taken back home. For example, returnees may come with metal roofing, nails, rice, and processed food. As freshly caught fish fade from view, the need for purchased tinned fish increases, and a newly organised world appears with disempowered villagers leaving destroyed environments for urban opportunity and uncertainty.[5] Nietschmann (1979) describes such an unfolding in the Caribbean.

The Value of Time

Both traditional harvesting of building materials from the local environment, and the purchase

Figure 13.1

Figure 13.2

Figure 13.3

Figure 13.4

Figure 13.5

Figure 13.6

Figure 13.7

Figure 13.8

Figure 13.9

Figure 13.10

Figure 13.11

Figure 13.12

of non-sustainable cash-required building materials, take time. Walking into the forest, for example, and seeking straight small trees of a particular species for felling; cutting and hauling them to the village; and debarking, shaping, and sizing these logs into small rafters or floor joists takes time. Later, after a number of years, these materials may begin to rot and more time will be invested to remove them, find new ones, and replace them.

For a non-sustainable material such as sheet metal roofing, the time involved in, say, fishing, transporting fish to market, making the return trip, and subsequent iterations can be a considerable investment of time by villagers. Then, with inexperience as there is no traditional usage, the metal is affixed to the traditional house. It is often considered that such a material is worth the time because it does not require the maintenance that a natural roof of, say, sago palm needs. Or is this the case (Figure 13.14)?[6]

Time is an important consideration in village life. If the reason for purchasing sheet metal for roofing and siding is to never have to maintain the roof and walls again, then the occupants have estimated that there will be less time expenditures in the future. Any possible costs incurred to the environment or excessive amounts of time involved in gaining the nonsustainable and presumed durable building material will be justified by future time saved.

Village time has not slipped totally into the "time is money" equation of those enmeshed in the cash economy. Time, in the fully sustainable Papua New Guinean village, is instead valued as a limited asset to be carefully budgeted for such important tasks as gardening, housebuilding and maintenance, yard- and house-keeping, fishing and hunting, community and cultural affairs, and so on. These variations are of course related to the dual economies of cash and sustainability, and greater and lesser degrees are present in many locales. Nevertheless the relationship of time to building materials and the implied promises of future saved time through the use of nonsustainable building materials may be misleading.

Heading for Analyses

In the village, uneasily situated at the crossroads of sustainability and the global economy, a number of questions must be asked and answered by the villagers if they are to make decisions that safeguard their environment, their village, and their way of life. Rhetorical questions can set the stage. How much village land is used for sustainable building, and how many people can the land sustain, and is the village population nearing that level? How much time does it take to build sustainably and how does this compare to the time spent acquiring cash by harvesting village resources for nonsustainable materials? What sustainable materials are used and for what? What are the levels of harvesting by the villagers, and the estimated regeneration rates? How can selective logging occur for village income? How many eels should be trapped and sold at the market and still have enough for the village and the river? To what extent can village gardens be market focused? These and other questions address the issue of how villagers can knowledgeably protect their tribal holdings and themselves.

Figure 13.13

Figure 13.14

Such questions, put into operationalised hypotheses, result in data in both quantitative and qualitative forms. The complexity of these data does not lend itself to simple analysis. But in matrix and graphic form the data can be seen, discussed and decisions made based on knowledge (Miles and Huberman, 1984). This is a preferred alternative to present forms of village decision-making which may rely on unfounded assumptions and opinions. And it is through graphic material that data can most effectively be presented to villagers. Subsequent discussions and decisions can readily be driven by the villagers, with or without aid from outsiders.

Participating in the Sustainable Village Life and the Global Economy: Picking and Choosing

An investigation of a traditional Papua New Guinean village house will uncover many examples of fastening materials together by lashing with sustainable rope-like materials such as tree bark, inner fibers from plants, and vines. Small girts might be lashed to rafters and might have sago roofing panels yet lashed to them. Where timber sizes are large enough, most villagers today gladly avoid lashing and use, instead, steel wire nails.[7] Nails are not too costly, and they create sturdy joints if properly used in a braced structure. They are very efficient with regard to time when compared to lashing (Figure 13.3).

This may well be an example of an appropriate use of a product of the global economy. It may take a few trips to market and a few extra rows in the garden, but the percent of the garden used to purchase nails will likely be modest and not cut into needed food for the family. Of course, knowing how to use nails is a prerequisite for a secure shelter and points out that new materials are not effective if the builder does not understand correct usage of a new technology (Figure 13.15).

The ridge cap on traditional coastal roofs is woven coconut palms (Figure 13.5). Today, more and more commonly, one finds this cap (which is susceptible to occasional water leaks) replaced by a small amount of bent sheet metal roofing. The cost is not great and the benefits of dryness real. Villagers can and do choose building materials and with the aid of data from studies can pick those materials whose costs do not endanger the sustainable life.

Comfort and Nonsustainable Materials

A purchased metal roof put on a traditional village house form transforms the intricately worked out dynamics of human comfort into chaos and the need arises to readapt, to fine tune, and eventually to redesign the house. The metal roof does not function as does the sustainable sago or pandanus roof which allows the rising heated air to slip through the small interstices of the thatch. Rather, hot air accumulates underneath the impervious metal and converts the house into a hot box (Figure 16). Where the thinness of the metal radiates the sun's heat into the house, the thatched roof has a thickness created by a multitude of entrapped air spaces between fibrous fronds and is a form of insulation against heat.

Understanding human comfort in the traditional house and in newer hybrid houses

Figure 13.15

Figure 13.16

requires a knowledge of the factors at work. Human response (in terms of heat stress) to the thermal environment depends primarily on four environmental variables: air temperature, air movement, radiation (mean radiant temperature), and humidity. Personal variables include: acclimatisation, body shape, subcutaneous fat, age, and gender. Indoor thermal comfort in a warm humid climate depends largely on the control of radiant heat and the promotion of air movement (cooling breezes) in a building. The control of radiant heat requires the consideration of massive versus light-weight construction, roof design, orientation of the building, and solar heating. Whilst the promotion of air movement requires the consideration of orientation of the building, house form and plan, elevation of the building, and the use of verandahs and pavilions.

In addition to the above, it must be noted that thermal comfort is usually difficult to define in precise terms as it depends on personal preferences including thermal sensation as well as psychological and social factors. An attempt, however, at estimating the thermal comfort zone can be done by measuring environmental variables such as air temperature, air movement, and humidity. Other methods of estimation can include the measurement of indoor surface temperatures of ceilings, walls, and floors and comparing these with outdoor variables such as incoming solar radiation, outdoor air temperature, outdoor relative humidity, and outdoor air movement and direction. These methods use experimental techniques involving appropriate instrumentation.

Tests involving instrumentation are not currently possible due to the unavailability of such devices to the researchers. Instead, a thermal performance study using the techniques of observation and interviewing will be conducted with rural villagers. It is hoped that interview sessions together with physical observations will highlight the advantages and disadvantages of the use of both sustainable and nonsustainable building materials.

Recycling

With or without a knowledge of how different building materials perform, villagers are apt to incorporate them into their houses. They know, by and large, that many of the materials of the global economy are firm and durable and can withstand time better than the organic sustainable materials of the village, and they are inclined to accept these materials in lieu of traditional materials that require periodic renewal.

When the Papua New Guinean economy — now depressed — was more robust anecdotal evidence suggests that city refuse including castoff sheets of plywood, old plastic laminate kitchen cabinet counters, rusted metal roofs, and cardboard. These and other such detritus found their way into nearby village houses. And in villages more distant from urban centers, the piece of plywood found afloat at sea or the left-over material from a mining site, or the chance find during a city visit, also offered something useful to incorporate into the house (Figures 13.17, 13.18, 13.19).

There is an element of environmental conservation in such recycling, even though it might offend aesthetes. It makes economic good sense for the villager, and it extracts further use-

Figure 13.17

Figure 13.18

value from a product that others considered without value.

Now, however, the chances of acquiring such materials at no cost has diminished as economic times worsen because they are less likely to be jettisoned by economically constrained urbanites. And, for the aesthete, the person concerned about the ultimate disposal of unwieldy pieces of manufactured materials, and for the villager suffering from a house without the creature comforts common in traditional houses of sustainable materials, the absence of recycled materials is not without a note of pleasure. But there are other reasons why the great bulk of nonsustainable building materials, whether recycled or not, may be inappropriate for the Papua New Guinean house, and these are social and cultural.

The Traditional Flexibility of the Sustainable Village House and the Rigid Materials of the Global Economy

User transaction in built settings such as a family dwelling directly addresses two architectural components that facilitate spatial dwelling experience by the user-occupant. One of the components is the visually present physical building structure whilst the other addresses a more subtle entity that has a 'hidden dimension' (Hall 1966) of social architectural significance. Social architecture addresses and accommodates social and cultural issues that make up the hidden dwelling experience and their influence on the behavior and interactions of the user-occupants, and often determines the outcome of the end product, the dwelling

Figure 13.19

setting. Thus, the notion of dwelling is a socio-cultural experiential norm and fusion between the built structure and human behavioral interaction within the domestic dwelling sphere and its related peripheral-space relationships in many Melanesian indigenous societies. The human interface with the building is seen (Gonduan 1999) to have greater significance than the physical structural make-up of the building itself. However, there exists some measurable variables between physical architectural space and social architectural spatial developments for which further research is required among indigenous Melanesians.

Recent research (Gonduan 1999) carried out among indigenous Papua New Guineans indicates that there is a greater fit and/or congruence in traditionally built buildings and settings using traditional materials than in buildings and settings that are fabricated using nontraditional and nonsustainable materials and approaches to building construction. The research also suggests that, due to this fit, there is greater control (Reser 1979, Bell et al 1996), and manageability by the owner-builder, thus giving greater satisfaction to the user-occupant in traditionally built settings (Gonduan 1999), compared to those who have opted for conventional (e.g. modern National Housing Commission Low Cost Housing type) means as experienced in most urban and now increasingly in rural communities in Papua New Guinea (Figure 13.13).

Dwelling settings[8] utilising traditional methods of building and construction that incorporate locally harvested renewable building materials allow for flexibility not available when manufactured standardised and nonsustainable materials are used. This flexibility allows the user-occupants to have greater proximate, experiential choice and decision control over their own designs and dwelling settings than can be available with nontraditional, nonsustainable, and rigid material.

Though the user appears to have control over the building of nonsustainable dwelling types, he/she is limited and/or constrained by physical and economic dimensions that are outside of his/her control and in most cases dictate the overall outcome of the dwelling. Money thus often becomes an issue, and the user-occupant seeks cash in order to buy construction materials for the building. Often this becomes the major constraint, along with material rigidity and the physical limitations of factory-sized building material, and results in inflexibility in housing form. The structure may not actually meet owner-occupant housing needs. Whereas if the user-occupant utilised traditional local building materials and methods there would be minimal constraints on material use and shape resulting in greater flexibility to build the type of building that would comfortably accommodate socio-cultural fabrication of the overall and total dwelling experience.

For example, if a rural village man decides to build an extension to an existing house and/or build a new house as necessitated by social and/or cultural developments in his family, he would have to go into his family/clan/tribal forest and harvest all the sustainable building material required for the new construction. The easy accessibility to available building material allows the user-occupant to be flexible in making choices on how he is able to build. Such choice and flexibility can be achieved using nonsustainable building material only when large sums of money are expended and it is currently very difficult for ordinary rural Papua New Guineans to be financially rich.

In this perspective there is a need to establish a connection between the building which utilises traditional building materials, and building methods that accommodate social architectural flexibility. Such housing is seen as having a greater value to rural villagers when compared to those building types that utilise nonsustainable and alien material from the global economy. Moreover the house of alien materials, in that they do not promote and/or uphold the social and cultural values of sustainable living in rural areas, is seen as damaging to the social, cultural, economic, and environmental survival of indigenous community in Papua New Guinea and in the greater Melanesian Region.

Conclusion

It is our intent to develop a methodology of assessment that will be accessible to and usable by villagers in Papua New Guinea. The goal is to allow them to maintain their preferred life style; to guard their environment from over-exploitation; and to allow them to see the desires, forces, and pitfalls of the global economy and to selectively pick and choose as they see fit. We are hopeful that this descriptive piece will suggest to researchers in other locales the value of undertaking similar studies in their countries.

NOTES

1. See, for example, the results of over-harvesting by Miskito Indians on the Nicaraguan eastern coast (Nietschmann 1979).
2. While circular migration has been documented in Papua New Guinea as a prominent form (Skeldon 1977, Ross 1984), over extraction of environmental resources such as those discussed here, has the potential to impact migration and can inhibit return, see, for example, Curtain (1976) on the ratio of usable land for cash cropping opportunities to high population numbers and the absence of high returnee rates in a Sepik River village.
3. All photographs were taken by one of the authors in 1998, in Morobe and Milne Bay Provinces. Out of respect for villagers we refrain from citing village locations. Those whose legitimate research interests may require such identification should contact the authors.
4. We are indebted to Mr. Ken Polin, Lecturer at the University of Technology, for this observation made at his home village on Manus Island.
5. Historically this is not a new phenomenon. The rural people of England were forced into the cities during the Industrial Revolution when many hands were needed. The French peasant, unpleasantly situated with too many children, bundled them off to be city servants in the 1800s. And the small-acreage farmer in the United States, economically made insignificant by agri-business, lost heart and mortgage, and had to settle for city jobs when they could be found. This need not be the case in Papua New Guinea.

6. An argument of note is that coastal locations have salt-laden air currents that will so quickly rust out a metal roof that the life expectancy is not any greater than the five to seven years normally expected of a sustainable sago or pandanus roof. Research, however, is required for substantiation or refutation.

7. A concern that nailed joints expose the interior of the timber to moisture penetration and thus rot cannot be dismissed. A small study is warranted to explore this matter and to suggest how nails holes, if they lead to rot, can be sealed, covered, or otherwise protected against water intrusion.

8. 'Dwelling setting' refers to the total domestic dwelling parameter used and utilised by the dwelling user-occupant under social and cultural observations and practices, and is not necessarily limited to the physical dwelling structure as experienced in many Papua New Guinean and Melanesian communities.

REFERENCES

Altman, I., (1992) "A Transactional Perspective on Transitions to New Environments." *Environment and Behaviour*, 24(2).

Altman, I. and Setha, M. L., (1992) "Place Attachment." In *Human Behaviour and Environment, Advances in Theory and Research*, Vol. 12. London: Plenum Press.

Bell, P. A., Greene, T. C., Fisher, J. D., and Baum, A., (1996) *Environmental Psychology*. Sydney: Harcourt Brace.

Brehm, S. S. and Brehm, J. W., (1981). *Psychological Reactance: A theory of freedom and control*. New York: Academic Press.

Curtain, Richard, (1976). The 1974-75 Rural Survey. A Study of Outmigration from Fourteen Villages in the East Sepik Province. Boroko: Institute of Applied Social and Economic Research.

Gonduan, C. K., (1999). "Space Use and Space Utilisation in the Native Melanesian Domicile. A case study of the Iatmul people in the Middle Sepik River Region of Papua New Guinea," PhD thesis. Australian Institute of Tropical Architecture, James Cook University.

Hall, E. T., (1966) *Hidden Dimension*. Garden City, New York: Double-Day.

Miles, Matthew and A. Michael Huberman, (1984) *Qualitative Data Analysis: A Sourcebook of New Methods*. Beverly Hills: Sage Publications.

Nietschmann, Bernard, (1979) "Ecological Change, Inflation, and Migration in the Far Western Caribbean." *The Geographical Review*, 69 (1).

Skeldon, Ronald, (1977) Internal migration in Papua New Guinea: a statistical description. Boroko: Institute of Applied Social and Economic Research.

Reser, J., (1979) "A Matter of Control: Aboriginal Housing Circumstances in Remote Communities and Settlements". In *A Black Reality: Aboriginal camps and housing in remote Australia*. Edited by M. Heppell. Canberra: Australian Institute of Aboriginal Studies.

Ross, Anthony C., (1984) Migrants From Fifty Villages. Boroko: Institute of Applied Social and Economic Research.

Wallerstein, Immanuel, (1984) *Politics of the World Economy*. New York: Cambridge University Press.

Appendix

Book Review

The Internationalisation of Environmental Protection

Edited by Miranda A. Schreurs and Elizabeth C. Economy
Cambridge University Press, 1997; pp. xiv+224
ISBN: 0-521-58536-8

In the editors' selection of scholarly articles for *The Internationalization of Environmental Protection*, they acknowledge the advent of environmental problems into the global forum. Schreurs and Economy look at why there has become a need for collective measures to be taken at a global level to combat various environmental problems and they also discuss some of the factors that influence some nations in their domestic and international policy-making strategies. Both unilateral and multilateral policies have played reciprocal roles in negotiations and in forming international and national environmental protection institutions and policies. Seven articles (plus an introductory article by the editors themselves) deal with the variety of approaches taken by major players in global environmental policy-making.

The eight articles and the contributors are: "Domestic and International Linkages in Environmental Politics" by Elizabeth C. Economy and Miranda A. Schreurs (editors), "Chinese Policy-making and Global Climate Change: Two-front Diplomacy and the International Community" by Elizabeth C. Economy, "The Domestic Politics of Global Biodiversity Protection in the United Kingdom and the United States" by Kal Raustiala, "Domestic and International Linkages in Environmental Politics: A Case-study of the Montreal Protocol" by Joanne M. Kauffman, "The Internationalization of Environmental Protection in the USSR and its Successor States" by Robert G. Darst, "Domestic Institutions and International Agendas in Japan and Germany" by Miranda A. Schreurs "Zimbabwe and CITES: Illustrating the

Reciprocal Relationship Between the State and the International Regime" by Phyllis Mofson; and "The European Union: Bridging Domestic and International Environmental Policy-making" by Angela Liberatore. In general, the articles demonstrate how and why nations responded or did not respond to certain environmental issues and abatement measures negotiated at the global level. Some of the environmental issues touched upon were global climate change, global biodiversity protection, the depletion of the ozone layer and endangered species. The most impressive articles deal with China's case with the global climate change negotiations, Zimbabwe's case in negotiating the Convention on International Trade in Endangered Species of Wild Flora and Fauna (CITES) and the United States' reluctance to sign the Biodiversity Convention.

It is not surprising that the editors have selected articles that deal with the most powerful actors in the world such as the US, the UK, Japan, Germany, the former Soviet Union, China and the EU. Indeed, there is only one article that deals with a non-major actor in the international arena. Yet even in this case, the article shows how a less powerful nation, Zimbabwe, attempts to influence policy at a global level by aligning itself with a major power and also by joining forces with its fellow developing nations. For instance, Japan, as a consumer of ivory, sided with Zimbabwe, the producer state, in order to influence the outcomes of CITES negotiations so that the results were in their favour. However, as was observed, the attempt was a failure and Zimbabwe had to pursue an alternative approach.

The editors have done well in their selection of articles dealing with the international strategies of the most influential nations in global environmental policy-making. However, one would have hoped they could have shed more light on the efforts of nations outside the group of developed nations. When one picks up a book with the title *The Internationalization of Environmental Protection*, one would expect to see not only a consideration of the developed nations (which includes China as a rapidly developing nation), but also a wider and more balanced inclusion of developing nations' global environmental strategies. Failing to do this, one would expect the editors to provide more articles dealing with the impact on international environmental policy-making by developing nations.

As might be predicted, nations took an agreeing stance if they perceived that some benefit was coming from the decision they were about to make. If no benefit was forthcoming, they took the contrary stance or attempted an alternative approach. In other words, the action or inaction of a nation has been usually interest-oriented. These interest motivated policies are seen in the cases of both China and Zimbabwe who were concerned with domestic industry, trade and economic growth. Also, the United States' reluctance and lack of prompt response to sign and ratify the Biodiversity Convention was due to the consequences this would have had on its domestic legal regime and its workforce. Even though industry and other national affairs have become the primary factors that influence domestic action and shape global environmental policies and guidelines, it is also illustrated that multilateral actions or the international regime could also demand changes in domestic action, policies, and legal and political institutions. Zimbabwe's submission to the decision of the international community to ban ivory trade is one such example.

The book also explains the reciprocal effects of environmental policies. Sometimes nations acted unilaterally by proposing policies for multilateral negotiations and agreement. For instance, China's five principles, which won international support (especially from the developing nations), were successfully used throughout all of China's international negotiations and Zimbabwe's unsuccessful efforts to redirect international agreement on the ban of ivory trade exemplifies the reciprocal effect of unilateral and multilateral actions. Moreover, these articles pointed out that environmental pollution and degradation caused by one or a few nations often becomes a common problem for a region or the world as a whole. However, individual nations have not been capable of financing clean-ups of the pollution they created for themselves and others. Therefore, nations devoid of financial and other resources to combat their local environmental problems have acted collectively with others so that financial, technical and labour assistance would in turn be provided by the global community. The international community has become an avenue for gaining resources needed to tackle environmental problems and, in this way, the environmental problems have become internationalised. The former Soviet Union and its successor states and China are a few of the nations discussed in the articles that have used this tactic of global assistance.

On the whole, despite the unbalanced treatment of articles concerning developed nations, as well as little mention of the NGOs and the multinational corporations (with the exceptions of Du Pont and ICI) as part of international environmental policy shapers, the main concepts and factors which forge environmental policy-making both at the domestic and global levels are present in the book. This includes a thorough look at the significant factors in global and domestic environmental policy-making such as the concepts of unilateralism, multilateralism and interests, industries, institutions, domestic politics and economic situations. I believe this collection of works would be useful for students of international relations. Specifically, it is a good book for those intending to engage in the future with domestic or global environmental policy-making and relevant disciplines. The environmental issues, the different factors determining international environmental policy outcomes and the tactics used by various actors discussed in this book are worth knowing of in this time of changing trends in domestic and global environmental policy-making.

Alphonse Kambu

www.ingramcontent.com/pod-product-compliance
Lightning Source LLC
Jackson TN
JSHW062249110926
108091JS00011B/24
9789980945747